Anonymous

Log of the U.S. Gunboat Gloucester

Anonymous

Log of the U.S. Gunboat Gloucester

ISBN/EAN: 9783337017941

Printed in Europe, USA, Canada, Australia, Japan

Cover: Foto ©berggeist007 / pixelio.de

More available books at **www.hansebooks.com**

LOG

OF THE

U. S. Gunboat Gloucester

COMMANDED BY
LT.-COMMANDER RICHARD WAINWRIGHT

AND

THE OFFICIAL REPORTS

OF THE

PRINCIPAL EVENTS OF HER CRUISE DURING THE LATE

WAR WITH SPAIN

ANNAPOLIS, MD.
PUBLISHED BY PERMISSION OF THE NAVY DEPARTMENT
BY THE U. S. NAVAL INSTITUTE
1899

EXTRACT FROM THE REPORT OF REAR-ADMIRAL SAMPSON, COMMANDER-IN-CHIEF, ON THE BATTLE OF JULY 3, 1898.

The skilful handling and gallant fighting of the Gloucester excited the admiration of every one who witnessed it, and merits the commendation of the Navy Department. She is a fast and entirely unprotected auxiliary vessel—the yacht Corsair—and has a good battery of light rapid-fire guns. She was lying about two miles from the harbor entrance, to the southward and eastward, and immediately steamed in, opening fire upon the large ships. Anticipating the appearance of the Pluton and Furor, the Gloucester was slowed, thereby gaining more rapidly a high pressure of steam, and when the destroyers came out she steamed for them at full speed, and was able to close to short range, where her fire was accurate, deadly, and of great volume.

During this fight the Gloucester was under the fire of the Socapa battery. Within twenty minutes from the time they emerged from Santiago harbor the careers of the Furor and Pluton were ended and two-thirds of their people killed. The Furor was beached and sunk in the surf; the Pluton sank in deep water a few minutes later. The destroyers probably suffered much injury from the fire of the secondary batteries of the battleships Iowa, Indiana, and Texas, yet I think a very considerable factor in their speedy destruction was the fire, at close range, of the Gloucester's battery. After rescuing the survivors of the destroyers, the Gloucester did excellent service in landing and securing the crew of the Infanta Maria Teresa.

LOG BOOK

OF THE

U. S. S. _Gloucester_

___4th___ Rate,

of ___10___ Guns,

COMMANDED BY

Richard Wainwright, Lt. Comdr. U. S. Navy,

Attached to _North Atlantic_ Squadron,

Commencing _May 16_, 1898,

at _New York_,

and ending _September 4_, 1898,

at _New York._

LIST OF OFFICERS OF THE U. S. S. GLOUCESTER.

Name.	Duty.	Joined.	Detached.	Remarks.
Lieut.-Commander RICHARD WAINWRIGHT....	Captain	May 16	Nov. 2	U. S. Navy
Lieut. HARRY P. HUSE............	Executive and Navigator	" 16	Sept. 28	U. S. Navy
Lieut. THOMAS C. WOOD........	Watch and Division Officer.	" 22	Sept. 23	Hon. Dis.
Lieut. G. H. NORMAN, JR........	"	" 18	Sept. 28	Hon. Dis.
Ensign J. T. EDSON..............	"	" 16	Sept. 28	Hon. Dis.
Assistant Surgeon J. F. BRANSFORD.........	Promoted to Passed Assistant Surgeon..	" 16	Oct. 13	Hon. Dis.
Assistant Paymaster ALEXANDER BROWN......	Assistant Paymaster.......	" 16	Nov. 5	Hon. Dis.
Passed Assistant Engineer G. W. McELROY..........	Promoted to Chief Engineer..........	" 16	Oct. 19	U. S. Navy
Assistant Engineer A. M. PROCTER............	Watch and Division Officer.	" 16	... .	U. S. Navy

QUINTARD IRON WORKS, NEW YORK, *Monday*, May 16, 1898.
11 A. M. to midnight.

At eleven o'clock ante meridian the ship was formally put in commission. Officers present: Lieutenant-Commander Richard Wainwright, Captain; Lieutenant Harry P. Huse, Executive; Ensign John T. Edson; Assistant Surgeon J. F. Bransford; Assistant Paymaster Alexander Brown; Passed Assistant Engineer George W. McElroy. Assistant Engineer Andre M. Procter reported onboard for duty. Workmen from Quintard Iron Works onboard engaged in installing gun-mounts, caulking, painting, fitting battle screens for deck-houses, and generally overhauling interior of ship.

HARRY P. HUSE, Lieutenant, U.S.N., Executive and Navigator.

QUINTARD IRON WORKS, NEW YORK, *Tuesday*, May 17, 1898.
Midnight to midnight.

Moored to dock at foot of East 12th Street, New York. Workmen from Quintard Iron Works onboard installing gun-mounts, caulking, painting, etc.

HARRY P. HUSE, Lieutenant.

QUINTARD IRON WORKS, NEW YORK, *Wednesday*, May 18, 1898.
Midnight to midnight.

Same as yesterday. Lieutenant George H. Norman, U. S. N., reported onboard for duty.

HARRY P. HUSE, Lieutenant.

QUINTARD IRON WORKS, NEW YORK, *Thursday*, May 19, 1898.
Midnight to midnight.

At same place as in previous watch. Workmen from Quintard Iron Works onboard. Main engines were given a dock trial in the afternoon. Finished painting ship.

HARRY P. HUSE, Lieutenant.

NAVY YARD, NEW YORK, *Friday*, May 20, 1898.

Midnight to midnight.

Cast off from dock at Quintard Iron Works at two o'clock in the afternoon and proceeded to Navy Yard in tow of navy-yard tug. Tied up to coal dock at Navy Yard and, at four o'clock, received crew from U. S. Receiving Ship Vermont with papers and effects.

E. C. ADKINS	Fireman, 2d Class.
T. S. BARNES	Machinist, 2d Class.
C. BECHTOLD	Seaman.
A. BECKARY	Apothecary.
G. W. BEE	Electrician, 2d Class.
J. BOND	Seaman.
B. BOWIE	Machinist, 2d Class.
O. BROWN	Seaman.
A. R. CANNON	Ordinary Seaman.
M. J. CARABINE	Machinist, 2d Class.
G. H. CAREY	Mess Attendant.
C. CARLSSON	Seaman.
J. CARR	Cabin Cook.
G. CHIPMAN	Electrician, 2d Class.
T. COLLERAN	Fireman, 2d Class.
C. A. COLLIN	Seaman.
A. J. COLE	Coal Passer.
J. COOKSEY	Landsman.
H. J. COXE	Landsman.
H. DAHL	Seaman.
A. DUNCAN	Landsman.
M. K. DAVIS	Landsman.
H. ENGLERT	Seaman.
C. ENSMANGER	Landsman.
G. B. EVANS	Seaman.
L. E. GRAVES	Electrician, 2d Class.
H. C. GREEN	Seaman.
O. HALVERSON	Ordinary Seaman.
W. A. HARBOUR	Ordinary Seaman.
R. T. HARE	Machinist, 2d Class.
H. H. HILLMAN	Seaman.
J. HOULAHAN	Fireman, 2d Class.
E. HOWARD	Coal Passer.
R. JENNINGS	Machinist, 1st Class.
C. JOHANSON	Fireman, 1st Class.
C. JOHNSON	Shipwright.
S. KASTELL	Ordinary Seaman.
P. KAY	Landsman.
P. KELLER	Seaman.
L. KLEINKOPF	Landsman.
E. KRONING	Seaman.
F. W. LACY	Chief Yeoman.

THE CREW.

J. W. LEWIS..............................Ordinary Seaman.
B. LOEHRS ....Landsman.
H. H. LYKKE.............................Seaman Gunner.
P. LYNCH ....Seaman.
G. LYNN................................Cabin Steward.
T. MACKLIN.............................Ordinary Seaman.
S. L. MACNAIR.........................Ordinary Seaman.
M. MAGEE ....Landsman.
P. A. MEEHAN..........................Seaman.
M. MULCAHEY ....Seaman.
M. J. MURPHY ....Seaman.
W. A. McARTHUR ....Blacksmith.
W. H. McKEON ....Oiler.
J. McMILLEN ....Fireman, 1st Class.
H. McNAB..............................Fireman, 2d Class.
W. NELSEN ....Mess Attendant.
G. NOBLE ....Ordinary Seaman.
L. QUENTIN ....Landsman.
H. ROBERTS ....Fireman, 2d Class.
G. W. RILEY ....Ordinary Seaman.
G. RUDISCHHAUSER ....Fireman, 2d Class.
*J. P. SATLER ....Ordinary Seaman.
W. H. SELLERS ....Chief Yeoman.
W. J. SULLIVAN ....Coal Passer.
M. J. TIERNEY ....Seaman.
A. D. THOMPSON ....Seaman.
G. VENEAU ....Coal Passer.
P. WELCH ....Coal Passer.
S. WHETTON............................Landsman.
W. W. WHITELOCK......................Landsman.
J. WINTERS ....Shipwright.
F. WIRTANE............................Seaman.
G. E. WOODSIDE........................Machinist, 2d Class.
J. W. WILLIAMS........................Coal Passer.

Crew had supper on Vermont. Engaged in receiving stores.

HARRY P. HUSE.

NAVY YARD, NEW YORK, *Saturday*, May 21, 1898.
Midnight to midnight.

Moored to coal dock. Receiving stores in Equipment, Construction, Ordnance, and Steam Engineering Departments. Navy yard workmen onboard preparing to instal guns. Crew cleaning ship and receiving stores. Received from U. S. R. S. Vermont A. Jaggi, seaman, with papers and effects.

HARRY P. HUSE.

*See foot note on p. 14.

NAVY YARD, NEW YORK, *Sunday*, May 22, 1898.
Midnight to midnight.

Moored to coal dock. Receiving stores as in preceding watch. Absent without leave: John P. Satler (ordinary seaman).* Lieutenant Thomas C. Wood, U. S. N., reported onboard for duty.

HARRY P. HUSE.

NAVY YARD, NEW YORK, *Monday*, May 23, 1898.
Midnight to midnight.

Moored to coal dock. At 10 A. M. coal lighter came along starboard side. Commenced coaling immediately. Received 92 tons bituminous steaming coal. Navy yard workmen from Ordinance Department installing three-pounder and six-pounder rapid-fire guns. Receiving and stowing ammunition. Received the following men with bags, hammocks, and necessary papers: James Buchanan (mess attendant), Frank Butler (mess attendant), Richard Daly (ordinary seaman), William H. Lawrence (ordinary seaman), Walter F. Lee (coal passer), H. G. Pond (electrician, 2nd class), Charles Rozzle (ordinary seaman), from U. S. R. S. Vermont. At about 9.10 P. M. fire alarm was sounded from navy yard. Took crew with buckets, hose, and hand grenades to scene of fire, building No. 6. Fire under control at ten, when crew returned to ship.

HARRY P. HUSE.

---

* A mistake. Satler was on the detail for the Gloucester, and his papers were received, but he was never sent onboard.

Navy Yard, New York, *Tuesday*, May 24, 1898.
Midnight to noon.

Making preparation to get under way. Navy yard workmen finished installation of guns. Filled bunkers from lighter alongside.

HARRY P. HUSE.

Noon to 4 P. M.

Weather fair, becoming foggy; wind light from southeast; barometer falling. At noon cast off shore moorings and steamed down East River. At one o'clock came to off Stapleton in consequence of fog, and anchored in ten fathoms of water with thirty fathoms of chain on starboard anchor. Engaged in cleaning ship. Fog shut in thick at about three and lifted somewhat by four. Barometer at four, 29.96.

GEORGE H. NORMAN, JR.

4 to 6 P. M.

Weather foggy; light air from southeast. Sent gig's crew in whaleboat ashore with mail. Light rain.

J. T. EDSON.

6 to 8 P. M.

Weather foggy; wind light from southward and westward. Sundown at 7.12. Ship swung to flood at seven. At 6.45 U. S. S. Resolute came down from Navy Yard and anchored off New Brighton, Staten Island. Savannah steamer anchored abeam.

THOS. C. WOOD.

8 P. M. to midnight.

Weather foggy with light air from southeast; thicker weather by ten o'clock; freshening breeze. Less thick at twelve with southerly air.

GEORGE H. NORMAN, JR.

NEW YORK BAY, *Wednesday*, May 25, 1898.
Midnight to 4 A. M.

Weather foggy; little or no air stirring.

J. T. EDSON.

4 to 8 A. M.

Thick fog, thinning occasionally; calms and light airs from north-ward and eastward; ebb tide. At six o'clock sent second whale-boat with gig's crew ashore with mail; returned at seven.

THOS. C. WOOD.

8 A. M. to noon.

Weather foggy; very light breeze from southeast. At 8.30 breeze shifted to northwest.

GEORGE H. NORMAN, JR.

Noon to 4 P. M.

Weather foggy but less dense; light airs from the south and west. At two o'clock Lieutenant Wood was sent ashore in second whale-boat to take reciprocal bearings. Between two and three o'clock ship was swung through a few points, by turning over engines, for deviation. Paymaster serving out stores to 1st division on berth deck.

J. T. EDSON.

4 to 6 P. M.

Weather wet; light fog with occasional showers; light airs from south and east. Ship swinging to ebb tide. At 4.45, U. S. S. Free Lance, Lieutenant Hanus Commanding, arrived in and anchored on our port beam. At 5.50 second whaleboat was prepared to go ashore with mail.

THOS. C. WOOD.

6 to 8 P. M.

Rainy weather; watch uneventful.

GEORGE H. NORMAN, JR.

8 P. M. to midnight.

Light breeze from northeast; drizzling rain; mist; flood tide.

J. T. EDSON.

NEW YORK BAY AND AT SEA, *Thursday*, May 26, 1898.
Midnight to 4 A. M.

Wind fresh from northward and eastward; swung to ebb at 12.15; light rain and fog; wind dropping. At 3.15 clearing. Weather variable.

THOS. C. WOOD.

4 to 8 A. M.

Weather squally; wind east north east, freshening. At 5.40 got underway and passed through Main and Gedney channels. At 7.07 took magnetic course S 7° E from whistling buoy. At eight o'clock wind increased to moderate gale; heavy sea running. Patent log read 9.8 at whistling buoy.

GEORGE H. NORMAN, JR.

8 A. M. to noon.

Threatening weather; moderate gale from the east; rough sea; ship rolling badly so as to take in solid water through hawse holes. At 8.37 turned back and at nine o'clock made Scotland lightship. Came to anchor in lower bay west of Sandy Hook at ten. At time of turning back, patent log read 10.58.

J. T. EDSON.

Noon to 4 P. M.

Weather squally with rain; moderate gale continues from eastward. At anchor under lee of Sandy Hook; six fathoms of water; thirty fathoms of chain. Crew busy at routine duties.

THOS. C. WOOD.

4 to 6 P. M.

Weather squally with rain; moderate gale from eastward. Sent second whaleboat away with mail to ordnance dock at 4.40. At six o'clock boat was still away.

GEORGE H. NORMAN, JR.

6 to 8 P. M.

Whaleboat returned at seven. No change in weather.

J. T. EDSON.

8 P. M. to midnight.

Overcast and hazy; strong easterly wind with light rain during first part of watch.

THOS. C. WOOD.

2

AT SEA, *Friday*, May 27, 1898.

Midnight to 4 A. M.

Weather squally with occasional showers.

GEORGE H. NORMAN, JR.

4 to 8 A. M.

Overcast; occasional light rain; wind variable from north to east.

J. T. EDSON.

8 A. M. to noon.

Weather overcast; light airs from eastward and southward. At 10.15 called all hands up anchor. At 10.20 steamed ahead bound out. Passed Scotland lightship at 11.23; patent log reading 23.25; course by standard compass, south. At twelve o'clock patent log read 28.70.

THOS. C. WOOD.

Noon to 4 P. M.

Weather cloudy and overcast; light easterly winds. At 3.14 Barnegat light was abeam, five miles distant by bow and beam observations. Courses by standard compass: twelve to 1.15, south; 1.15 to 3.14, S ½ W; 3.14 to four, S by W ½ W.

GEORGE H. NORMAN, JR.

4 to 6 P. M.

Overcast and cloudy; light easterly breeze. Course S by W ½ W. At 4.30 soundings showed 10 fathoms; at 6.00 P. M. 7 fathoms. At six o'clock Absecom light was on starboard beam distant about eight miles.

J. T. EDSON.

6 to 8 P. M.

Weather overcast; light airs from southward. At six o'clock sounded and found 7 fathoms; patent log 92.5; Absecom light bore west north west a half west. At 6.30 changed course to S ¾ W by standard compass. At 6.30 sounded and found 7 fathoms. At 7.15 found 8 fathoms; patent log 4.6. Sunset at 7.15. Set running lights. At eight o'clock sounded and found 14.5 fathoms; patent log 12.2.

THOS. C. WOOD.

8 P. M. to midnight.

Weather overcast; wind south to west. At 9.15 with Five Fathom lightship abeam changed course to south, standard compass; log reading 23.3; sounding 17. At midnight barometer read 29.83; log 47.6.

GEORGE H. NORMAN, JR.

At Sea, *Saturday*, May 28, 1898.

Midnight to 4 A. M.

Overcast and cloudy; some haze; wind southwest. At 3.50 sighted a sail one point on starboard bow. This proved to be a four-masted schooner which passed to starboard of us. The soundings taken at 12.30, 1.30, 2.30, and 3.30 corresponded closely with those laid down on the chart. At 2.30 log read 67.9; soundings 27 fathoms. At 3.30 log 77.3; soundings 34 fathoms.

J. T. Edson.

4 to 8 A. M.

Weather overcast; light haze; wind light from south south west; sea smooth. At 4.40 took bearing of sun at rising N 66° E. At five o'clock sounding showed 45 fathoms; reported to Captain. At six sounding machine showed no bottom with 250 fathoms of line out and 90 fathoms by tube. At 6.05 Captain ordered course changed to south three quarters west. Weather clearing and, at eight, blue sky with bright sun. No bottom with 270 fathoms of line and 100 fathoms indicated by glass tube. All hands engaged in cleaning ship.

Thos. C. Wood.

8 A. M. to noon.

At eight o'clock, weather clear overhead with haze in horizon; light airs from southwest; sea smooth. At noon, blue sky with haze in horizon.

George H. Norman, Jr.

Noon to 4 P. M.

Weather clear; light haze; light airs from south by west. Soundings taken every hour without finding bottom, 185 fathoms of line out. At two o'clock, patent log read 71.2; at four o'clock, 86.3. Barometer 29.88.

J. T. Edson.

4 to 6 P. M.

Weather clear; wind light from south south west. Navigator engaged in swinging ship between four and five o'clock. At five put ship on her course south three-quarters west; patent log reading 86.4. At six o'clock changed course to south south west; patent log reading 96. Sounded and found no bottom at 98 fathoms with 185 fathoms of line out.

Thos. C. Wood.

6 to 8 P. M.

Weather clear with light south south west wind. At seven o'clock wind freshened. Heavy clouds in western horizon.

GEORGE H. NORMAN, JR.

8 P. M. to midnight.

Slight haze; light breeze from south south west. At nine o'clock log read 27; soundings 16.5 fathoms. At 10.30 sighted Hatteras light bearing west a half north. At eleven o'clock sighted Diamond Shoal lightship bearing south a half east. Turned lightship at 11.40 patent log reading 54.9; changed course to south south west three-quarters west.

J. T. EDSON.

AT SEA, *Sunday*, May 29, 1898.

Midnight to 4 A. M.

Weather clear. Dropped Diamond Shoal light astern at 12.30. At 12.35 passed a steamer bound north. Wind from southward and westward. Soundings at one o'clock, 35 fathoms; two o'clock, 24 fathoms; three o'clock, 24 fathoms; four o'clock, 22 fathoms.

THOS. C. WOOD.

4 to 8 A. M.

Weather clear with haze in horizon. Passed to port two steamers bound north, at too great distance to distinguish colors.

GEORGE H. NORMAN, JR.

8 A. M. to noon.

Weather clear. Sighted a steamer on starboard beam at 8.20. Changed course at 9.30 to southwest by south; log read at that time 46. Latitude by observation at noon 33° 37′ N.

J. T. EDSON.

Noon to 4 P. M.

Weather warm and pleasant; hazy near horizon; wind light from south south west. At one fifty sighted a schooner on starboard beam standing to southward and eastward. At two o'clock sighted a schooner one and a half points on port bow standing to northward and eastward. At 2.10 two sails were reported, one on port bow and one on port beam. At 3.40 a steamer was reported two points forward of starboard beam. No change in ship's course from previous watch.

THOS. C. WOOD.

4 to 6 P. M.

Weather clear. At 4.05 swung ship for compass compensation. At 5.14 set course S W by W ½ W.

GEORGE H. NORMAN, JR.

6 to 8 P. M.

Weather clear; wind southwest. Log at eight o'clock, 46; soundings, 17 fathoms.

J. T. EDSON.

8 P. M. to midnight.

Weather clear; wind from southward and westward. When sounding with sounding machine at nine o'clock lost overboard lead and depth recorder. No sails or lights seen during watch.

THOS. C. WOOD.

AT SEA, *Monday,* May 30, 1898.

Midnight to 4 A. M.

Clear sky with scattered cirrus clouds and light haze in horizon.

GEORGE H. NORMAN, JR.

4 to 8 A. M.

Sky blue and clear with a few cirrus clouds only; gentle breeze from southwest, and a moderate swell of the sea. No sail or light seen during the watch. Soundings, 10 to 13 fathoms.

J. T. EDSON.

8 A. M. to noon.

Weather clear and pleasant; wind light from southward and westward. At nine o'clock changed course by order of Commanding Officer to S S W ¾ W. At nine thirty exercised at general quarters At eleven o'clock sent all bedding on deck to air. —————— —————— (seaman) accused of absence from ship without leave while in New York was brought to the mast and made a prisoner at large by the Commanding Officer, awaiting trial by court-martial; in the meantime performing his duties as usual.

THOS. C. WOOD.

Noon to 4 P. M.

Weather fair with cirro-stratus clouds and hazy horizon. At two o'clock by order of Commanding Officer changed ship's course from S S W ¾ W to S by E.

GEORGE H. NORMAN, JR.

4 to 6 P. M.

Weather clear and pleasant. At evening quarters the guns' crews were drilled at target practice; two shots being fired from each gun; range from 300 to 900 yards. Most of the shots were well placed.

J. T. EDSON.

6 to 8 P. M.

Weather clear and pleasant; wind from southward and westward.

THOS. C. WOOD.

8 P. M. to midnight.

Weather clear; wind southwest.

GEORGE H. NORMAN, JR.

Midnight to 4 A. M.   AT SEA, *Tuesday*. May 31, 1898.

Weather clear and pleasant; gentle breeze from south south west. At 12.55 changed course to southeast; patent log reading 59.6. At 2.15 changed course to south by east; patent log, 73.8; soundings, 45 fathoms. At four o'clock no bottom found with 75 fathoms of line.

4 to 8 A. M.   J. T. EDSON.

Weather clear and pleasant; wind light from west south west. At five o'clock lookout reported a light bearing two points on starboard bow. Took soundings every hour of the watch, using glass tubes, and found no bottom at 50 fathoms. Course unchanged, south by east.   THOS. C. WOOD.

8 A. M. to noon.

Weather clear; sky cloudless; light haze on horizon. Changed course at 8.30 to south; patent log reading 44.2. Changed course at 9.04 to south south west; patent log reading 51.3. Changed course at 11.13 to south: patent log reading 76. ——— ——— (ordinary seaman) was awarded one week's extra duty for being dirty.

Noon to 4 P. M.   GEORGE H. NORMAN, JR.

Clear and pleasant; slight haze near horizon. Clothing was served out by Paymaster. Hammocks aired. Boats overhauled.

4 to 6 P. M.   J. T. EDSON.

Weather clear and pleasant; hazy near horizon. At 4.20 sighted a steamer's smoke three points on the port bow. At 4.25 a sail was sighted on the starboard bow. At 5.40 land was visible with strong glasses on the starboard side from two points forward of the beam to two points on the bow. At evening muster served out clean hammocks and passed the word for all hands to scrub hammocks at six o'clock.   THOS. C. WOOD.

6 to 8 P. M.

Weather clear. At 7.40 sighted Jupiter Inlet light. At eight o'clock the light bore two and a half points on our starboard bow.

8 P. M. to midnight.   GEORGE H. NORMAN, JR.

Jupiter light was dropped on starboard quarter at ten o'clock. Changed course to south by east a half east at 9.15; log 73; lead 11 fathoms. At 10.24 changed course to south; log 82.5; lead 28. At 11.07 changed course to south a quarter west; log 89.3; lead 71. At 11.30 changed course to south a half west; log 92; lead 75.

        J. T. EDSON.

AT SEA, *Wednesday*, June 1, 1898.

Midnight to 4 A. M.

Warm and pleasant; at 3 o'clock changed course to S by W, patent log reading 20.4. Soundings as follows were registered by sounding machine: At 12.30, 83 fathoms; at 1.00, 25 fathoms; at 1.20, 20 fathoms; at 1.45, 30 fathoms. As the spring recorder failed to record properly, since 170 fathoms of wire was run out, at 2.30 used glass tubes, which showed 100 fathoms of water with 195 fathoms of line out. At three o'clock no bottom was found with 200 fathoms of line out. No lights or vessels seen during watch. At four o'clock no bottom was found with 200 fathoms of line out.

THOS. C. WOOD.

4 to 8 A. M.

Weather clear; hazy horizon. At 4.55 changed course to southwest; at 5.20 changed course to south; at 5.30 changed course to S ½ W, patent log reading 42.1, in 5 fathoms of water. At 6.29, with Biscayne buoy abeam, changed course to south, patent log reading 50.2, in 9 fathoms of water. At 6.34 changed course to S ½ E. Passed close under bows of schooner Cowes G. Kerwin at anchor, standing by wrecked schooner half a mile to the westward. At 7.07 had Florida buoy abeam, patent log reading 56.3. At 7.35 Old Tower was abeam, patent log reading 60.2.

GEORGE H. NORMAN, JR.

8 A. M. to noon.

Weather clear: slight haze. At 8.42 changed course to S by W ¼ W, patent log reading 72.9. At nine o'clock found no bottom at 73 fathoms. At 10.10 found bottom at 66 fathoms; changed course to S S W ¼ W. At 10.30 patent log read 87.5; depth of water 59 fathoms. At 10.40 changed course to southwest. At 10.50 Turtle Harbor buoy bore W S W, patent log reading 90.3. At 10.55 changed course to S W ½ S. At 11.05 changed course to S S W ½ W, patent log 92. At 9.30 sighted a steamer one point on the starboard bow; all hands were called to quarters, and as the steamer showed no colors a solid shot was fired across her bow, whereupon she ran up the English flag and put her helm to starboard. We signalled with three blasts of the whistle and proceeded on our course. At 11.55 Carysfort light-house bore abeam (W N W ½ W), distant about one mile, patent log reading 97.3. At 12.33 with Elbow Beacon abeam, log reading 3.4, changed course to S W ¾ S. Grecian Shoals were abeam at 12.51; patent log 5.3.

J. T. EDSON.

Noon to 4 P. M.

Weather pleasant. At 1.40 French Reef bore abeam, patent log reading 12.3. At 1.40 a steamer was sighted on port bow steering to southward and eastward. At 1.55 changed course to S W ½ W, patent log reading 14.4. At 2.10 changed course to S W by W ¼ W, patent log not recorded. At 2.12 changed course to W S W; at 2.40 changed course to S W ¾ W. At 3.50 o'clock passed Alligator Reef light bearing abeam, distant 12 miles. Taking soundings by machine every hour.

4 to 6 P. M.

Weather pleasant; changed course at 4.12 to S W ¼ S; at 4.24 to S W; and at 5.18 to W S W ¾ W. At 7.10 Sombrero light was abeam, 2.3 miles distant by bow and beam bearings.

GEORGE H. NORMAN, JR.

6 to 8 P. M.

At 8.05 flash-light of American Shoal was two points on the starboard bow; it bore west a half south at 8.14.

J. T. EDSON.

8 P. M. to midnight.

Weather pleasant. At 9.55 ran in near American Shoal light, and anchored in 20 fathoms of water, with 60 fathoms of chain on starboard anchor.

THOS. C. WOOD.

KEY WEST, FLORIDA, *Thursday*, June 2, 1898.

Midnight to 4 A. M.

Weather clear with light mist on horizon at midnight, becoming thicker towards morning. Ship lay heading east with American Shoal light bearing west.

GEORGE H. NORMAN, JR.

4 to 8 A. M.

Weather clear and pleasant. Got up anchor at 5.30 A. M., and lay course W S W. Beacon A abeam at 6.20. Sand Key light-house was one point on starboard bow at 6.45. Changed course to W ¾ S at 6.30, and to W ¼ N at 7.10. The harbor of Key West in sight on starboard bow; several vessels at anchor and alongside the wharves.

J. T. EDSON.

8 A. M. to 8 P. M.

Anchored off Government landing at Key West about nine o'clock. Ship swinging to ebb tide. At eleven o'clock water lighter came alongside. In the afternoon sent liberty parties ashore and made arrangements for coaling ship.

THOS. C. WOOD.

8 P. M. to midnight.

Weather clear; moon nearly full. At 8.30 tide turned and began to ebb, all the vessels in the stream turning about with it; at 11.30 wind freshened.

J. T. EDSON.

KEY WEST, FLORIDA, *Friday*, June 3, 1898.

Midnight to 4 A. M.

Lying at anchor; ship swung to flood tide at 2.30 A. M. Light airs from eastward.

J. T. EDSON.

4 to 8 A. M.

At anchor off Key West; light easterly airs: riding to flood.

J. T. EDSON.

8 A. M. to noon.

Lying at anchor coaling; ship swung to ebb tide at eight o'clock.

J. T. EDSON.

Noon to 4 P. M.

Swung to flood tide at four o'clock.

GEORGE H. NORMAN, JR.

4 to 8 P. M.

At four twenty having sent away, nearly empty, the barge from which we had been coaling we were set alongside the coal schooner Rebecca Landen by Government tug Nezinscot with our starboard to her port side and proceeded with our coaling.

GEORGE H. NORMAN, JR.

8 P. M. to midnight.

Lying at anchor; swung to tide at ten o'clock. Weather clear and pleasant; light breeze from eastward.

J. T. EDSON.

KEY WEST, FLORIDA, *Saturday*, June 4, 1898.

Midnight to 4 A. M.

Weather pleasant; ship swung to flood tide at 3.15.

GEORGE H. NORMAN, JR.

4 to 8 A. M.

Called all hands at 4. 30; turned to at five o'clock.

8 A. M. to noon.

Swung to ebb tide at 8.30. Naval Cadet S. W. Bryant and Lieutenant A. H. Dutton reported on board with orders. The Montgomery, Marietta, Rodgers and Dupont came to anchor in the harbor at 7.30. Received on board Colonel Hernandez (Cuban Insurgent Army) and two pilots as passengers. Got underway and stood out through ship channel under one boiler.

J. T. EDSON.

Noon to 4 P. M.

Clear and fine; wind easterly, moderate breeze. No sails sighted during watch. Course S E by E. Patent log at end of watch 20.8.

A. H. DUTTON.

4 to 6 P. M.

Fresh breeze from east north east. Sighted a sailing vessel at 4.40 three points on the starboard bow. At five o'clock sighted another sailing vessel two points on the port bow. Ship's infantry company drilled from 5.30 to 6.00 P. M. At 5.40 connected second boiler. Sail sighted one point forward of the port beam at 5.45.

THOS. C. WOOD.

6 to 8 P. M.

Breeze continues fresh; changed course to southeast at six o'clock.

GEORGE H. NORMAN, JR.

8 P. M. to midnight.

Fresh breeze from east; at nine o'clock saw a water spout ahead which passed to south of us. A slight squall struck us shortly after. At 10.10 changed course to E S E ½ E, patent log 81. At 10.20 made out a flashlight bearing S ¼ W; interval between flashes one minute. Light bore S W by S at eleven o'clock; patent log 89.4.

J. T. EDSON.

Midnight to 4 A. M.   At SEA, *Sunday*, June 5, 1898.

Bright moonlight; breezes from E S E, diminishing in force: halo around moon; vivid lightning to northward. Course E S E ½ E; distance run by patent log during watch 44.9 miles. Patent log at end of watch 45.6.

A. H. DUTTON.

4 to 8 A. M.

Weather clear and pleasant; daylight at 4.30. At 4.45 sighted a steamer hull down four points on starboard bow standing west north west. At 6.15 land was reported on starboard beam. At 7.05 changed course to S E by E ½ E.

THOS. C. WOOD.

8 A. M. to noon.

Distant mountains and nearer headlands abeam and on starboard bow. Lighthouse at 9.20 bore 49° forward of beam; patent log reading 3.9. At 10.32 lighthouse was abeam, patent log reading 16.9. At 10.45 changed course to S E by E, patent log reading 19.2. Set clock ahead nine minutes at noon. At 12.04 made Lobos Cay lighthouse two points on port bow.

GEORGE H. NORMAN, JR.

Noon to 4 P. M.

At 12.30 sighted a steamer ahead; sent crew to general quarters. Signals were exchanged and vessel proved to be the U. S. S. Mayflower. At 1.45 Cay Confite was about one point forward of the starboard beam and Cayo Verde one point abaft the starboard beam. Lobos Cay lighthouse dropped out of sight on port quarter at half past two.

J. T. EDSON.

4 to 6 P. M.

Rain squalls from southeast and north during entire watch. Lighthouse sighted on starboard bow at 4.20, dropped out of sight about six o'clock. Sighted a sail on port bow at 5.45. At 4.35 changed course to S E by E ¾ E; patent log read 76.6.

A. H. DUTTON.

6 to 8 P. M.

Weather cloudy, light showers from northward and eastward. Lightning around horizon.

THOS. C. WOOD.

8 P. M. to midnight.

Moonlight with mackerel sky. At 8.35 changed course to E S E ¾ E, log reading 8.9, with instructions to run this course at 8 knots per hour for 60 miles.

GEORGE H. NORMAN, JR.

AT SEA, *Monday*, June 6, 1898.

Midnight to 4 A. M.

Weather pleasant; light air from east and south. Course E S E ¾ E.

J. T. EDSON.

4 to 8 A. M.

Fair, warm and pleasant; light to moderate breezes from south eastward. At 4.25 changed course to S by E. At 5.05 made land ahead and on starboard bow. At 5.20 changed course to S ¾ W; at 6.07 to S W; at 7.05 to W ½ N. Standing in to harbor of Banes. Sighted a small sloop close inshore near entrance to Nipe harbor at 7.15. Took cast with sounding machine at 7.55, 120 fathoms of line out; no bottom.

A. H. DUTTON.

8 A. M. to noon.

Standing in to Port Banes. Entrance to harbor very narrow and tortuous. Assisted by Cuban pilots made inner harbor without difficulty and anchored off Port Banes at 10.30 in five fathoms of water. Sent boat ashore with Colonel Hernandez and pilots and received visits from natives with fruit.

THOS. C. WOOD.

Noon to 4 P. M.

Riding to our starboard anchor off railroad pier, Banes, with fifteen fathoms of chain in five fathoms of water. Tide was running ebb at noon and had fallen twelve inches and was still falling at four o'clock.

GEORGE H. NORMAN, JR.

4 to 8 P. M.

At anchor in four fathoms at four o'clock. Shifted anchorage at five o'clock into five fathoms of water; ship swinging to flood tide. Ship stopped swinging at 5.55. Sent armed boat's crew under Lieutenant Dutton to guard entrance of harbor against attack from Nipe.

GEORGE H. NORMAN, JR.

8 P. M. to midnight.

At anchor in seven fathoms of water with fifteen fathoms of chain out. Early in the evening, sky covered with thunder and rain clouds; at midnight, bright moonlight.

GEORGE H. NORMAN, JR.

BANES, CUBA, *Tuesday*, June 7, 1898.

Midnight to 4 A. M.

At anchor in six fathoms of water; ebb tide, water fell about sixteen inches during the watch.

J. T. EDSON.

4 to 8 A. M.

Got up anchor off Banes at 6.30 and commenced working out through passage against strong flood tide. In turning first bend found it necessary to get out anchor astern; fouled buoy rope with screw; cleared it in thirty minutes and worked ahead.

THOS. C. WOOD.

8 A. M. to noon.

Weather pleasant; working our way out through passage from Banes harbor. Parted starboard chain and lost twenty-five fathoms of chain and the starboard anchor in swinging ship. Reached channel's mouth at 10.20, came to anchor and called in boat party sent out the previous evening to guard channel entrance. At 11.50 got underway and layed course E ¼ S, patent log reading 3.4.

GEORGE H. NORMAN, JR.

Noon to 4 P. M.

Weather pleasant; changed course at one o'clock to E ½ S.

J. T. EDSON.

4 to 6 P. M.

Slight swell, sky overcast and cloudy; at 4.22 patent log read 50 and course was changed from E ½ S to S E ½ E.

GEORGE H. NORMAN, JR.

6 to 8 P. M.

Overcast first part; clearing after seven o'clock. At 6.55 sighted Cape Maysi on starboard bow, and at 7.05 changed course to S E by E. Small white light sighted on starboard beam at 7.25.

A. H. DUTTON.

8 P. M. to midnight.

Weather cloudy first part, clearing latter part of watch. At 8.46 changed course to S ½ W, patent log 96.8. At 10.23, having run 18 miles, changed course to W S W ½ W, patent log 14.8.

THOS. C. WOOD.

SANTIAGO BLOCKADE, *Wednesday*, June 8, 1898.

Midnight to 4 A. M.

Weather sultry; moonlight. At 3.15 changed course to W ¾ N.

GEORGE H. NORMAN, JR.

4 to 8 A. M.

Weather sultry; light air off shore. Eastern point of Guantanamo bore N E ½ E at 4.45 A. M. Land sighted on starboard beam at daylight; changed course to N W ¼ W at 6.45, patent log 97.9. A fleet of vessels sighted at 7.30 bearing N W by W.

J. T. EDSON.

8 A. M. to noon.

Weather clear and hot; swell from southward and eastward. Made out vessels sighted in previous watch to be U. S. Blockading Fleet off Santiago under Rear-Admiral Sampson. Reported arrival to Flagship and delivered mail to fleet. Naval Cadet Bryant left the ship and reported for duty on board the Massachusetts.

THOS. C. WOOD.

Noon to 4 P. M.

Engaged in delivering mail to fleet. Weather sultry; nothing of interest occurred during watch.

A. H. DUTTON.

4 to 8 P. M.

Weather muggy but clear. About five o'clock received orders from Flagship to keep close by for orders; at six got orders to land a Cuban party at Acerradero. At 6.30 received party and ran down to the Suwanee then to the Brooklyn and asked for a pilot. Were sent by her in search of one to the Vixen.

GEORGE H. NORMAN, JR.

8 P. M. to midnight.

Cloudy first part; clearing after ten, with bright moonlight. Prepared to land a Cuban general, with staff of six officers, at Acerradero.

A. H. DUTTON.

SANTIAGO BLOCKADE, *Thursday*, June 9, 1898.

Midnight to 4 A. M.

Weather clear. At 12.10 sent away armed boat's crew to land Cuban party at Acerradero three miles distant bearing north. Then ran off shore and at two o'clock ran in and picked up boat which had carried out her mission. Steamed east to our position in the blockade passing through outer line unchallenged.

GEORGE H. NORMAN, JR.

4 to 8 A. M.

Ship lying off Morro Castle, distant two miles, bearing north. Weather pleasant; swell from southeast.

J. T. EDSON.

8 A. M. to noon.

Weather pleasant; swell heavy from southeast. Lying off and on Morro.

THOS. C. WOOD.

Noon to 4 P. M.

Off Morro Castle; weather clear and pleasant; cross swells.

GEORGE H. NORMAN, JR.

4 to 6 P. M.

Clear and pleasant. At 5.30 Brooklyn signalled "close in to three miles at dark." ——— ———, ordinary seaman, was confined in double irons, for insubordination, by order of the Captain.

J. T. EDSON.

6 to 8 P. M.

Weather clear and pleasant. Lying off Morro. Closed in about eight o'clock to within one mile of entrance to Santiago harbor.

THOS. C. WOOD.

8 P. M. to midnight.

Clear and pleasant; starlight; bright moonlight after eleven. On station from one to two miles off Morro. Lights seen in harbor from time to time, upon approaching entrance closely.

A. H. DUTTON.

SANTIAGO BLOCKADE, *Friday*, June 10, 1898.

Midnight to 4 A. M.

Shore line distinctly visible. Morro Castle, distant two miles, bears north north east. Moonlight; weather pleasant. Five vessels of our fleet visible. Ground swell from southeast.

J. T. EDSON.

4 to 8 A. M.

Weather pleasant; daylight at 4.45. Ship laying off and on by Morro. At daylight took position about two miles to westward of Morro, and one and a half miles from shore.

THOS. C. WOOD.

8 A. M. to noon.

Weather pleasant. Received letters from the Brooklyn for the Commanding Officers of Massachusetts and flagship New York which we delivered. We also received our mail from the Armeria. Got orders from Brooklyn to stand by in attendance.

GEORGE H. NORMAN, JR.

Noon to 4 P. M.

Lying to off Morro, two and a half miles from shore. Weather clear; wind southwest, light; ground swell from southeast. Received a Cuban major and a pilot from Flagship New York at 3.30 P. M.

THOS. C. WOOD.

4 to 6 P. M.

Got underway by orders from Flagship and carried to Acerradero (18 miles to westward) a Cuban major and a pilot, and landed them with Dr. Bransford in charge at 5.30 P. M.

THOS. C. WOOD.

6 to 8 P. M.

Weather pleasant; lying off Acerradero for return of second whaler from shore in command of Dr. Bransford. Heavy ground swell rolling in; wind light from northwest.

GEORGE H. NORMAN, JR.

8 P. M. to midnight.

Bright starlight; on station about one mile from Morro at end of watch.

A. H. DUTTON.

SANTIAGO BLOCKADE, *Saturday*, June 11, 1898.

Midnight to 4 A. M.

Weather pleasant; moon rose about 1.30. Lying close inshore under Morro Castle watching entrance for enemy. Picket launches, Porter, and New Orleans in sight during part of watch. About three o'clock exchanged private night signals with battleship lying off shore.

THOS. C. WOOD.

4 to 8 A. M.

Clear weather with hazy horizon in early morning. Lying to two miles south of Morro, Santiago.

GEORGE H. NORMAN, JR.

8 A. M. to noon.

Left station off Morro at 8.20 and steamed slowly towards flagship New York. Spoke Flagship at nine o'clock, and were told to direct Scorpion to report alongside New York. Returned to station on blockade at 11.50. Morro bearing N W by N, distant four miles.

J. T. EDSON.

Noon to 4 P. M.

On blockade station off Morro Castle; weather clear, wind southeast, light. Received orders from Flagship to "come within hailing distance." Received orders and delivered same on board the Iowa and Oregon.

THOS. C. WOOD.

4 to 6 P. M.

Delivered orders and mail to Massachusetts and Brooklyn, and returned to station at 5.45. Morro bearing N 20° W, distant about four miles. Weather overcast and hazy. At 6.50 ran in towards night station.

GEORGE H. NORMAN, JR.

6 to 8 P. M.

Weather overcast and cloudy. Morro northwest, distant two miles.

J. T. EDSON.

8 P. M. to midnight.

Dark and overcast, with much lightning; heavy squall at 8.30. Took station, Morro bearing northwest from one-half to one mile distant. Fleet closed in on mouth of harbor. Moon rose about 11.30.

A. H. DUTTON.

SANTIAGO BLOCKADE, *Sunday*, June 12, 1898.

Midnight to 4 A. M.

Weather overcast; moon obscured by clouds. On inner blockade line; about one mile southeast from Morro. No lights allowed and no temperature or barometric observations taken.

GEORGE H. NORMAN, JR.

4 to 8 A. M.

Off Morro, distant one mile, bearing north a half west. Left for station near New York at five o'clock. Overcast and cloudy. Signalled by New York to come within hail. Uniform white dress.

J. T. EDSON.

8 A. M. to noon.

Weather warm and pleasant. At eight o'clock were ordered alongside Flagship; received instructions to proceed to Acerradero, obtain information regarding enemy's movements and disposition from Cuban officers, and report. Proceeded at once; reached Acerradero at 9.45; sent in first whaleboat with Dr. Bransford in charge. Lying off and on Acerradero.

THOS. C. WOOD.

Noon to 4 P. M.

At 12.40 whaleboat returned from shore with Doctor and Cuban major. Left Acerradero at 12.45; arrived alongside Flagship at 2.55 P. M.; sent wherry to Flagship with Captain, Doctor, and Cuban major. Weather pleasant, smoky, light breeze from east north east.

GEORGE H. NORMAN, JR.

4 to 6 P. M.

Stood out towards station off Morro at four o'clock. Weather pleasant.

J. T. EDSON.

6 to 8 P. M.

Weather unchanged from previous watch. As darkness came on, moved in to our night station under Morro Castle.

THOS. C. WOOD.

8 P. M. to midnight.

Clear and pleasant; bright starlight. Light breeze from northward. On station, one-half to one mile off Morro. At 10.30 heard several volleys of small arm fire at mouth of harbor.

A. H. DUTTON.

SANTIAGO BLOCKADE, *Monday*, June 13, 1898.

Midnight to 4 A. M.

Weather pleasant. Lying off Morro Castle, distant one mile north by west. One picket launch just inside of us. The Iowa and Oregon alternately approached our position and then receded. Some firing heard ashore.

J. T. EDSON.

4 to 8 A. M.

Weather pleasant; wind light from northward and eastward. Lying in close under Morro Castle until daylight, then moved slowly out and resumed day station near Flagship. Watch scrubbed clothes and mattress covers.

THOS. C. WOOD.

8 A. M. to noon.

Weather pleasant. At 9.45 Solace and St. Louis came in. At 11.30 were called to Flagship and received message for New Orleans.

GEORGE H. NORMAN, JR.

Noon to 4 P. M.

On blockading station off Morro. Weather smoky with rain squalls; strong breeze from southeast, declining in force towards the end of the watch. Delivered orders in wigwag from Admiral to the New Orleans to preserve her distance of four miles. St. Paul joined fleet at four o'clock and fired a salute of thirteen guns. Returned to our station.

A. H. DUTTON.

4 to 6 P. M.

Weather pleasant. Exercised infantry company at drill. At 6.30 were called alongside Flagship and ordered to communicate orders to Captain Folger of New Orleans to shell reported emplacement at daylight tomorrow and were notified that Vesuvius would bombard Morro during night.

THOS. C. WOOD.

6 to 8 P. M.

Clear and hot. Took night position in blockade at 7.45 of one-half to one mile south from Morro.

GEORGE H. NORMAN, JR.

8 P. M. to midnight.

Clear and pleasant; gentle breeze from north. On station, Morro bearing north, distant from one-half to one mile during watch. At 11.15, two shots were heard near mouth of harbor, apparently from field pieces, succeeded by a louder report, and then by a still louder one, coming from a vivid explosion, apparently of great force, about half way to summit of bluff west of Morro, and south of Cañones Point. Went to general quarters, and piped down at 11.35.

A. H. DUTTON.

SANTIAGO BLOCKADE, *Tuesday*, June 14, 1898.

Midnight to 4 A. M.

Weather clear and pleasant. Morro Castle north, distant one mile. Railroad fortifications east north east, distant two miles. Light air from northward. Lying off and on watching entrance to harbor.

THOS. C. WOOD.

4 to 8 A. M.

Weather clear. Left night station at five o'clock, and proceeded towards day station. At 5.30 New Orleans opened on battery on bluff one-half mile to the eastward of Morro with her starboard guns, then swung and used her port guns. Withdrew after ten minutes' firing. Fire was answered by battery and by a few guns in Morro and by some guns on western headland. One shell passed over and fell beyond the Gloucester.

GEORGE H. NORMAN, JR.

8 A. M. to noon.

Lying near Flagship. Received mail for the fleet. ——— ——— (landsman) for failing to hail boat in hostile waters was awarded five days double irons by Captain.

J. T. EDSON.

Noon to 4 P. M.

Weather clear and dry; wind southeasterly, fresh. Delivered mail to the Texas, Brooklyn, Scorpion, Iowa, Armeria, and put mail and pilot on board the Vixen. At 1.30 P. M. received (repeated) signal from the Brooklyn to come within hailing distance. Received orders from Flagship to proceed to Guantanamo.

THOS. C. WOOD.

4 to 6 P. M.

Weather bright and clear; fresh easterly wind. Proceeding at full speed to westward. Courses: at four, E; 4.05, E by S; 4.30, E ¼ N; 6.35, E by N ¼ N; 6.50, E by N. At 5.20 went to quarters and spoke the Porter. At six o'clock overhauled and passed the tug Dandy of Darien. At six, headland was abeam, patent log reading 72.8.

GEORGE H. NORMAN, JR.

6 to 8 P. M.

Weather pleasant; course E ¼ N. The harbor of Guantanamo was made at seven o'clock. Several vessels were there at anchor. Spoke the Marblehead at 7.40. Captain Wainwright went on board. The Armeria came to anchor near by at eight o'clock.

J. T. Edson.

8 P. M. to midnight.

Clear and pleasant; bright starlight. At 8.10 anchored at Guantanamo. U. S. S. Marblehead, Dolphin, Armeria, Scindia, Stirling and some other colliers in port. Commanding Officer visited Commander B. H. McCalla, U. S. N., senior officer present. At 11.20 got underway and steered course for Mole St. Nicholas. ——— ——— (landsman) was put in double irons for five days, by order of Commanding Officer, to serve out a sentence. Mr. Merryweather came on board for passage to Nicholas Mole.

A. H. Dutton.

AT SEA, *Wednesday*, June 15, 1898.

Midnight to 4 A. M.

Weather clear. At two o'clock spoke New York *Sun* steamer Eli, also headed for the Mole.

GEORGE H. NORMAN, JR.

4 to 8 A. M.

Weather clear and pleasant; moonlight. At 4.30 sighted a schooner standing to southward and westward. At five o'clock sighted a steamer on our starboard quarter standing to northward and westward. Made land ahead at sunrise. Changed course to E by N ¼ N at 6.15 A. M., patent log reading 44.1.

J. T. EDSON.

8 A. M. to noon.

Weather pleasant. Standing in towards St. Nicholas Mole, Haiti. Anchored off Mole at 10.35 A. M. in 13 fathoms of water, 45 fathoms on port chain. Sent Lieutenant Norman ashore in dingy with Mr. Merryweather (Herald correspondent), and steward.

THOS. C. WOOD.

Noon to 4 P. M.

At anchor in the harbor of St. Nicholas Mole, Haiti; strong breeze from northeast; weather clear. Sent first whaleboat ashore for Mr. Norman; boat returned at 11.20. Steamers Premier and Ravensdale, both British, arrived in harbor. Sent whaleboat to the Ravensdale, steward in charge; returned at 3.50.

HARRY P. HUSE.

4 to 8 P. M.

Weather pleasant; strong breeze from northeast. Put out drift lead. Served out Paymaster's stores to 1st and 2nd divisions. Acting appointments to date from May 21 were delivered to petty officers at evening muster as follows:

P. A. MEEHAN.. ....Seaman,	rated	Chief Master at Arms.
J. BOND.................. "	"	"	Boatswain's Mate.
C. BECHTOLD........ .... "	"	"	Quartermaster.
W. W. WHITELOCK......Landsman,	"	"	Yeoman.
J. WINTERS..............Shipwright,	"	"	Carpenter's Mate,
R. JENNINGS............Machinist, 1st Class,	"	"	Machinist.

P. KELLER	Seaman,	rated	Boatswain's Mate, 1st Class.	
G. B. EVANS	"	"	Quartermaster,	" "
H. C. GREEN	"	"	"	" "
G. W. Bee	Electrician, 2d Class,	"	Electrician,	" "
G. CHIPMAN	"	"	"	" "
R. T. HARE	Machinist, 2d Class,	"	Machinist,	" "
C. JOHANSON	Fireman, 1st Class,	"	"	" "
P. KAY	Landsman,	"	Ship's Cook,	" "
C. CARLSSON	Seaman,	"	Boatswain's Mate, 2d	"
E. KRONING	"	"	Quartermaster,	" "
G. RUDISCHHAUSER	Fireman, 2d Class,	"	Machinist,	" "
W. H. McKEON	Oiler,	"	Watertender.	
H. DAHL	Seaman,	"	Coxswain.	
A. JAGGI	"	"	"	
P. LYNCH	"	"		
G. NOBLE	Ordinay Seaman,	"	Quartermaster, 3rd Class.	
H. ENGLERT	Seaman,	"	Gunner's Mate,	" "
M. J. MURPHY	"	"	" "	" "
C. JOHNSON	Shipwright,	"	Carpenter's Mate. "	"
M. MAGEE	Landsman,	"	Painter.	

The following changes in rates were ordered by the Captain to date from May 21:

H. McNAB	Fireman, 2d Class,	rated	Fireman, 1st Class.	
H. ROBERTS	"	"	" "	
A. COLE	Coal Passer,	"	" 2d	"
H. J. COXE	Landsman,	"	Bayman.	

Sun dispatch boat got up anchor and put to sea at 6.15.

J. T. EDSON.

6 to 8 P. M.

Clear and pleasant; light winds from eastward. Sent Lieutenant Wood ashore in first whaleboat to receive dispatches.

A. H. DUTTON.

8 P. M. to midnight.

Clear and pleasant; bright starlight; light airs and breezes from east north east to southeast and calms. Lieutenant Wood returned with whaleboat at 8.20. Anchored with port anchor in 13 fathoms of water; 45 fathoms of chain at the hawsepipe.

A. H. DUTTON.

MOLE ST. NICHOLAS, HAITI, *Thursday*. June 16, 1898.
Midnight to 4 A. M.

Weather pleasant; light airs from south south east. At anchor off St. Nicholas Mole.

J. T. EDSON.

4 to 8 A. M.

Weather calm and pleasant. Daylight about 4.50. Called all hands at five o'clock. Scrubbed clothes, gun covers, and boat gear. At 7.25 sent second dinghy ashore in charge of Whitelock for dispatches.

THOS. C. WOOD.

8 A. M. to noon.

At 8.30 dinghy recalled with Whitelock missing. Sent first whaler ashore with Lieutenant Norman. Returned with information about Marcel's non-arrival, and with Whitelock at 10.05. Got immediately underway and at 10.20, on course W ½ N, passed Cape St. Nicholas abeam, patent log reading 82.8. Bearing dispatches. At 11.20 sighted American three-masted schooner broad off starboard bow bound west.

GEORGE H. NORMAN, JR.

Noon to 4 P. M.

Underway from St. Nicholas to Guantanamo, Cuba. Weather clear; wind northeasterly, strong breeze with swell. Sighted Guanas Point, Cuba, at one o'clock, bearing N W ½ N, distant twelve miles. Sighted steamer at two o'clock inshore running east. At three, off Galeta Point, ran off course in chase of steamer; fired a shot across her bows and brought her to. Sent Lieutenant Dutton to board her. On returning he reported she was the British steamer Jason of London, Captain Jas. Fraser, from Annatto Bay and Jamaican ports bound for New York. Papers regular; cargo, fruits, rum, island produce and ship's stores. Proceeded on course W S W ½ W.

HARRY P. HUSE.

4 to 6 P. M.

Weather pleasant; wind and swell from eastward. Running in for Guantanamo.

J. T. EDSON.

6 to 8 P. M.

Weather pleasant; wind strong from eastward with heavy swell. Bound to Guantanamo.

THOS. C. WOOD.

8 P. M. to midnight.

Clear and pleasant; light to moderate breezes from east north east; squally from eleven to midnight. At 8.10 came to anchor in outer harbor of Guantanamo, in 7 fathoms of water, 30 fathoms of chain at hawsepipe, inside. U. S. S. St. Paul at anchor on starboard bow; U. S. S. Solace on starboard beam; other U. S. vessels in inner harbor. Commanding Officer visited Captain C. D. Sigsbee, U. S. N., senior officer present.

A. H. DUTTON.

GUANTANAMO, *Friday*, June 17, 1898.

Midnight to 4 A. M.

Weather clear and warm. At anchor in Guantanamo Bay. At 1.30 hailed by St. Paul to keep a bright lookout on harbor entrance. From two to four St. Paul used her search-light on shore of bay.

THOS. C. WOOD.

4 to 8 A. M.

At 5.30 were directed by Captain McCalla to get our coal from Sterling, and by six were alongside the collier, and coaling.

GEORGE H. NORMAN, JR.

8 A. M. to noon.

Weather pleasant. Coaling ship from Sterling, which is alongside. By order of Captain, ———— ————, coal passer, was placed in double irons to await trial by court martial for leaving his station without permission, and ———— ————, landsman, was released from irons after three days' confinement. At 10.40 Vesuvius stood out to sea. New York *Sun* dispatch boat left at 11.30. One hundred and thirty-one buckets of coal received by 11.50. Lieutenant Dutton left the ship and reported for duty on the Marblehead. Marblehead and Suwanee shelled the northwest shore.

J. T. EDSON.

Noon to 4 P. M.

Weather pleasant. Coaling ship from U. S. S. Sterling; from two to four took in one hundred and fifty-four buckets, or about forty tons.

THOS. C. WOOD.

4 to 6 P. M.

Knocked off coaling at 5.30, having taken aboard about ninety-three tons. Oregon threw several shells into and towards Caymanera.

GEORGE H. NORMAN, JR.

6 to 8 P. M.

Got underway at 6.30 P. M. At 7.25 laid course southwest, patent log reading 87.5. At eight changed course to west, patent log 93.7. Weather pleasant.

J. T. EDSON.

8 P. M. to midnight.

Standing to the westward to rejoin fleet. Arrived off Santiago harbor about eleven o'clock; made night signal and ship's number; and took up night station off Morro.

THOS. C. WOOD.

AT SEA, *Saturday*, June 18, 1898.

Midnight to 4 A. M.

On night station off Morro. Weather pleasant. Watch uneventful.

GEORGE H. NORMAN, JR.

4 to 8 A. M.

Captain Wainwright went aboard Flagship at 5.15. After his return at 5.50 course was layed E S E ¼ E, patent log reading 24.1. Changed course to E ¼ N at 6.45, patent log 33.1. At 7.45 ran in towards a merchant steamer on port bow, bearing about E N E three miles away. At eight o'clock resumed course E ¼ N. Weather clear and pleasant.

J. T. EDSON.

8 A. M. to noon.

Weather pleasant. Engaged in cleaning ship. Course E.½N. —————— ——————, ordinary seaman, for inattention to duty was awarded five mid watches by Captain.

THOS. C. WOOD.

Noon to 4 P. M.

Underway from Santiago to St. Nicholas Mole; weather clear; wind east north east, light breezes. At 2.20 sighted two steamers nearly ahead and sheered off course to intercept them. At 2.40 passed U. S. S. Dixie, convoying U. S. S. Celtic, bound west. Exchanged numbers. Sighted land two points on starboard bow. At 2.45 P. M. resumed course E ¼ N.

GEORGE H. NORMAN, JR.

4 to 6 P. M.

Weather clear and pleasant; light breeze from east north east. Course changed to E by S at 3.30. Patent log hauled in at 5.12 reading 59.5. Came to anchor at St. Nicholas Mole at 5.40 in twenty fathoms of water.

THOS. C. WOOD.

6 to 8 P. M.

Weather warm and pleasant. At anchor at Mole St. Nicholas. At 6.30 tug Triton entered harbor.

THOS. C. WOOD.

8 P. M. to midnight.

At anchor at the Mole. At 8.30 cable steamer Marcel left the harbor followed a half hour later by tug Triton.

GEORGE H. NORMAN, JR.

AT SEA, *Sunday*, June 19, 1898.

Midnight to 4 A. M.

Clear and pleasant. At anchor in St. Nicholas Mole harbor.

J. T. EDSON.

4 to 8 A. M.

Clear and pleasant. U. S. collier Alexander came in.

THOS. C. WOOD.

8 A. M. to noon.

U. S. collier Alexander left port at 8.20. At 8.40 weighed anchor and passed out. With St. Nicholas Mole abeam at nine o'clock patent log read 59.7. Sighted steamer on our starboard bow at ten; bore down on her and at 10.20 spoke her, the Ely, flying the English flag. Hove to alongside the New York *Sun* tug; then stood away again for Guantanamo.

GEORGE H. NORMAN, JR.

Noon to 4 P. M.

Course west; making about thirteen knots. Weather clear and pleasant; light airs from west north west. Land in sight ahead and on starboard bow.

J. T. EDSON.

4 to 6 P. M.

Ran into Guantanamo and communicated with Commander McCalla of the Marblehead; then proceeded towards Santiago. At 4.40 took in patent log on rounding point at entrance to Guantanamo. At 5.22 put out log on standing away on course west, reading 57.2.

GEORGE H. NORMAN, JR.

6 to 8 P. M.

At 6.05 sent boat to tug Somers M. Smith and took dispatches for Admiral and others. Stood away on our course at 6.10, patent log reading 63.2. At 7.30 ——— ———, machinist first class, being intoxicated was put in double irons and, ——— ———, for neglect of duty as cook was given, in penalty, extra duty through midwatch for five nights, by Captain's order.

GEORGE H. NORMAN, JR.

8 P. M. to midnight.

Lying at usual night station off Morro Castle. Light airs from off shore.

J. T. EDSON.

SANTIAGO BLOCKADE, *Monday*, June 20, 1898.

Midnight to 4 A. M.

Weather clear and warm; wind light from northward and westward. Lying off Morro Castle, Santiago, on our night station.

THOS. C. WOOD.

4 to 8 A. M.

Weather pleasant. At 7.30 ran over to Resolute for mail.

GEORGE H. NORMAN, JR.

8 A. M. to noon.

Captain Chadwick came on board at 8.30; dinghy returned with mail. Set course southeast at 8.46; patent log 92.3. Sighted the Army fleet of forty sails about 10.10. Spoke Indiana at 10.25. Captain Chadwick went on board Segurança at 10.40, since which time we have kept her company, as she makes her way to the flagship New York.

J. T. EDSON.

Noon to 4 P. M.

Weather clear and pleasant; wind southeast, light. Followed steamer Segurança with General Shafter and staff. Admiral Sampson boarded her at 12.30 and proceeded to Acerradero. Admiral, General and staff, boarded us in the gig and after remaining some time proceeded in boats to the landing place (Acerradero).

THOS. C. WOOD.

4 to 6 P. M.

Lying off Acerradero, convoying the Segurança. At about 5.30 Vixen came up from the fleet and passed inside of the reefs. At six o'clock the Eagle came up flying the dispatch flag.

GEORGE H. NORMAN, JR.

6 to 8 P. M.

Lying off Acerradero.

J. T. EDSON.

8 P. M. to midnight.

Weather pleasant. Underway with Segurança and heading for the fleet off Santiago. At nine o'clock left Segurança and took our night position with the fleet off Morro.

THOS. C. WOOD.

SANTIAGO BLOCKADE, *Tuesday*, June 21, 1898.

Midnight to 4 P. M.

On night blockade station one-half mile southeast of Morro. Weather clear and watch uneventful.

GEORGE H. NORMAN, JR.

4 to 8 A. M.

Signalled from New York to come within hail at daylight. At 5.25 set course W N W ½ W. Ran towards reefs with pilot, and came to anchor at Acerradero, inside reef, at 7.50 A. M. Weather clear and pleasant.

J. T. EDSON.

8 A. M. to noon.

Weather pleasant. Lying inside reef off Acerradero. Took on board 299 Cuban troops. At eleven o'clock got underway and stood to southward, following Vixen.

THOS. C. WOOD.

Noon to 4 P. M.

Standing to the eastward accompanied by Vixen; heavy rolling sea. Three hundred Cuban soldiers lying seasick in heaps on deck forward. Stench intolerable.

GEORGE H. NORMAN, JR.

4 to 6 P. M.

Standing to eastward with Cuban soldiers on deck. Attempted to land Cubans, but were unable to do so on account of surf. Proceeded a few miles further to the eastward, following Vixen, to Sagua, and anchored there close to breakwater and in dangerous proximity to rocks to leeward.

J. T. EDSON.

6 to 8 P. M.

Weather pleasant. At anchor inside breakwater at Sagua. Landed Cuban troops (299).

THOS. C. WOOD.

8 P. M. to midnight.

Landed last of Cubans at 8.40; weighed anchor at nine o'clock and stood away for fleet off Morro. Crew went to supper at nine o'clock. Washed down decks and out lights by ten. Spoke Flagship, and received written orders at eleven; then proceeded to night blockading station.

GEORGE H. NORMAN, JR.

4

SANTIAGO BLOCKADE, *Wednesday*, June 22, 1898.

Midnight to 4 A. M.

Lying off night station; weather pleasant, light airs from off shore.

J. T. EDSON.

4 to 8 A. M.

Weather pleasant; wind light. At daybreak stood in to the position assigned to us under general orders of June 22, followed by Eagle, and waited for signal to begin bombardment of Aguadores.

THOS. C. WOOD.

8 A. M. to noon.

Weather clear. About 8.30, when off Aguadores, several shells were fired at Morro by guns of second division. At nine o'clock began shelling fortifications, earth works, bridges, and houses at Aguadores. Continued rapid firing with Eagle for fifteen minutes and then with slackened fire kept up bombardment till eleven. Few of the enemy were seen and no loss by them was observed. Our fire was not returned. Crew secured good gun practice and in many instances showed excellent markmanship.

GEORGE H. NORMAN, JR.

Noon to 4 P. M.

Lying off Aguadores; weather clear, light airs. At 1.30 P. M. shelled blockhouse at eastern end of bridge.

J. T. EDSON.

4 to 8 P. M.

Received mail from Flagship and stood in to our station.

THOS. C. WOOD.

8 P. M. to midnight.

Off Morro Castle. At 2.30 A. M. Vesuvius passed close under our stern, and shortly after sent three bombs which exploded to the east of Morro. Weather clear and pleasant.

J. T. EDSON.

SANTIAGO BLOCKADE, *Thursday*, June 23, 1898.

Midnight to 4 A. M.

Occupying our station off Morro and keeping watch on harbor entrance.

THOS. C. WOOD.

4 to 8 A. M.

On night blockading station under Morro; withdrew at daylight.

GEORGE H. NORMAN, JR.

8 A. M. to noon.

At 8.30 sent boat to Celtic for provisions by order from Flagship. Clear and pleasant.

J. T. EDSON.

Noon to 4 P. M.

Got two boatloads of ice from Celtic. Signalled from Flagship to come alongside. Received orders for Commanding Officer to go aboard.

THOS. C. WOOD.

4 to 6 P. M.

Commanding Officer returned at 4.10 with dispatches for Captain Goodrich. Stood away for the St. Louis at 4.12, patent log reading 59. Delivered dispatches at 4.55, log reading 68. Stood back for Flagship at 5.03, search-light beam bearing W ¼ S. Reached Flagship at 5.40, patent log reading 75.

GEORGE H. NORMAN, JR.

6 to 8 P. M.

Left Flagship for Acerradero at 6.50; course W by N; patent log 75; Morro bearing north by west.

J. T. EDSON.

8 P. M. to midnight.

Arrived off Acerradero at 8.20; sent boat with message to General Ludlow, U. S. Engineers, on board Alamo. Received his report in return and proceeded back to fleet reporting to Flagship at twelve o'clock. Weather squally with rain.

THOS. C. WOOD.

SANTIAGO BLOCKADE, *Friday*, June 24, 1898.

Midnight to 4 A. M.

Weather damp; sky partly overcast. At 12.30 the Commanding Officer having returned from Flagship we stood away at dead slow speed on a course E S E and at four o'clock had Daiquiri one point abaft the beam, distant about four miles.

GEORGE H. NORMAN, JR.

4 to 8 A. M.

Sent wherry to St. Louis at daybreak with orders from the Admiral to turn over one, or if possible, two steam launches for delivery to Flagship. At 7.15 left for Flagship, with two launches in tow. Weather close and sultry; light shore air.

J. T. EDSON.

8 A. M. to noon.

Commanding Officer repaired on board Flagship. On his return proceeded with two steam launches (New York and Massachusetts) to Acerradero. Summary court martial met for trial of —— ——.

THOS. C. WOOD.

Noon to 4 P. M.

At 12.10 court took recess till three o'clock. Hove to by transport Alamo off Acerradero at one o'clock and sent Lieutenant Wood aboard in launch of U. S. S. New York with messages for General Ludlow. On Lieutenant Wood's return, coaled and watered launch and, leaving her and the launch of U. S. S. Massachusetts in charge of Cadet Theleen with instructions to report to General Ludlow, we shaped our course for return to Flagship at 12.15. Off Daiquiri at four o'clock. Flagship bearing E ¼ S, distant six miles, patent log reading 11.6. Speed rate for one hour and thirty-five minutes 12.54 miles per hour.

GEORGE H. NORMAN, JR.

4 to 6 P. M.

Spoke Flagship and returned to station off Morro. Weather pleasant.

J. T. EDSON.

6 to 8 P. M.

Moved in to night position off Morro at dark.

THOS. C. WOOD.

8 P. M. to midnight.

On blockade at night station. Moonlight through hazy clouds. Watch uneventful.

GEORGE H. NORMAN, JR.

SANTIAGO BLOCKADE, *Saturday*, June 25. 1898.

Midnight to 4 A. M.

Lying off Morro at station; weather clear and pleasant.

J. T. EDSON.

4 to 8 A. M.

Weather warm and pleasant. In answer to signal from Flagship went alongside and reported as to situation at Aguadores. Received orders to return and prevent enemy from working on fortifications or railroad.

THOS. C. WOOD.

8 A. M. to noon.

Went over to New Orleans with message and made reconnaissance of Aguadores, returning to Flagship at nine o'clock. Carried orders and information to New Orleans and took position off Aguadores. Shelled earthwork and railroad cut when work was being carried on, driving enemy to shelter. Fired a few shells later when groups of men were seen. Anchored in nine fathoms of water distant 1000 yards from bridge and fortifications.

GEORGE H. NORMAN. JR.

Noon to 4 P. M.

After crew's dinner, took up sub-calibre practice with points on shore and the enemy for targets. Enemy then opened on us from fort with rapid fire machine gun and rifles, several shots passing through deck houses and upper works, but without doing injury to anyone aboard. Drove them from their position, battering down some battlements and silencing their fire. Weighed anchor. Were visited by Suwanee. Later left position and reconnoitered to the eastward. Returned at four o'clock.

GEORGE H. NORMAN, JR.

4 to 6 P. M.

No change in weather. Anchored off Aguadores. Observed no movements of enemy ashore. Took angles to determine position.

THOS. C. WOOD.

6 to 8 P. M.

Weather pleasant; light haze in sky overhead; moonlight. At anchor in nine fathoms of water, Aguadores fort bearing north, distant 900 yards. Dropped a few shells along the shore.

GEORGE H. NORMAN, JR.

8 P. M. to midnight.

Ship lying at anchor within 1000 yards of Aguadores. Bright moonlight until 10.30 P. M. At eleven o'clock turned search-light on shore, but no activity of any kind was observed. Deep sea lead carried away at eleven, caused by irregular and rocky bottom.

J. T. EDSON.

SANTIAGO BLOCKADE, *Sunday*, June 26, 1898.

Midnight to 4 A. M.

Weather pleasant. At anchor off Aguadores. Watch quiet. Could see no movement of enemy on shore.

THOS. C. WOOD.

4 to 8 A. M.

Weighed anchor at 4.15 and lay close in. Not a soul to be seen ashore and no change from yesterday in appearance of fortifications, earthworks, bridges, etc. At seven o'clock drifted off to southward; at 7.30 stood in again.

GEORGE H. NORMAN, JR.

8 A. M. to noon.

Off Aguadores; weather clear and pleasant: light breeze off shore. Came to anchor at nine o'clock in twelve fathoms of water; Spanish flag bearing north by west, distant 1800 yards. Went to general quarters at 9.30. Ship and crew were inspected by Lieutenant Huse, Executive Officer. Following punishment was assigned by Captain: —— ——, ordinary seaman, dirty and indifferent, five days' extra duty.

J. T. EDSON.

Noon to 4 P. M.

Weather clear and warm; wind southeast, light breeze. Riding to starboard anchor off Aguadores, distant 1300 yards. Punishment assigned by Captain: —— ——, landsman, going below when on watch and turning in, three days in double irons.

THOS. C. WOOD.

4 to 6 P. M.

Weather pleasant; wind southeast. Riding to starboard anchor: Aguadores fortress bearing north north west, distant 1300 yards. Heavy cumulus and rain clouds to southward at six o'clock.

GEORGE H. NORMAN, JR.

6 to 8 P. M.

Weather pleasant. At 7.45 P. M. a light was seen on shore high up to the eastward. One shot from gun No. 5 was fired at it.

J. T. EDSON.

8 P. M. to midnight.

Weather variable; squalls with rain from eastward. At 8.15 Dupont brought orders to leave Aguadores and take up our night station. Got up anchor at once and moved to our position off Morro. —— ——, landsman, was confined in double irons by order of Commanding Officer to carry out a sentence.

THOS. C. WOOD.

SANTIAGO BLOCKADE, *Monday*, June 27, 1898.

Midnight to 4 A. M.

Overcast; on night blockading station; watch uneventful.

GEORGE H. NORMAN, JR.

4 to 8 A. M.

At six o'clock it was observed that the western end of railway bridge at Aguadores had been destroyed during the night. The Flagship was informed by wigwag that northwest end of railroad bridge was blown up at 3 A. M. Steamship Yale came in from eastward at daylight.

J. T. EDSON.

8 A. M. to noon.

Off Aguadores. Signalled from Vixen and from Flagship that our pickets were approaching from eastward. Were cautioned not to mistake them for the enemy. Engaged in painting awnings and in other ship work.

THOS. C. WOOD.

Noon to 4 P. M.

Off Aguadores. Weather pleasant. At 2.30 men were observed working westward along the shore a mile to the eastward. Drew in close to the eastern shore of Aguadores light and by three o'clock could see many men in the bush and along the railroad track advancing towards the bridge. They displayed a Cuban flag to us and stopped their advance when within a few hundred yards of the eastern end of the bridge. We fired a few shells into the beach entrenchments under the fort where the enemy could be seen gathering, and got no response.

GEORGE H. NORMAN, JR.

4 to 6 P. M.

Off Aguadores; weather pleasant.

J. T. EDSON.

6 to 8 P. M.

Weather pleasant. At dark left Aguadores and stood over to our night station off Morro.

THOS. C. WOOD.

8 P. M. to midnight.

On night station in blockade, Morro bearing northwest to north north west, distant three-quarters of a mile. Moonlight; light air from northward. Watch uneventful.

GEORGE H. NORMAN, JR.

SANTIAGO BLOCKADE, *Tuesday*, June 28, 1898.

Midnight to 4 A. M.

Lying off Morro on night station; weather clear and pleasant.

J. T. EDSON.

4 to 8 A. M.

Lying at our station off Morro; Morro bearing about northwest. At daylight moved over to Aguadores. Observed smoke from Cuban camp to eastward of bridge.

THOS. C. WOOD.

8 A. M. to noon.

Off Aguadores. Cubans to be seen in bush and sheltered positions to the eastward of creek; enemy in rifle pits on beach below fort. At 10.30 o'clock fired a shell into a group by iron house and one into mouth of cave under fort.

GEORGE H. NORMAN, JR.

Noon to 4 P. M.

Wind northerly; weather squally; light breeze between squalls. Lying off Aguadores, distant 2200 yards, bearing north north east. Discovered several men in the bushes to the eastward of Aguadores bridge.

J. T. EDSON.

4 to 6 P. M.

Lying off Aguadores; weather pleasant. Swell heavy from southward and eastward.

THOS. C. WOOD

6 to 8 P. M.

Off Aguadores. Between seven and 7.30 heard considerable volley firing on shore and some artillery. Fired several times with bow gun. Took night station in blockade shortly before eight o'clock.

GEORGE H. NORMAN, JR.

8 P. M. to midnight.

Off Morro; weather pleasant. A picket launch lying close in was fired on about 10.30 P. M.

J. T. EDSON.

SANTIAGO BLOCKADE, *Wednesday*, June 29, 1898.

Midnight to 4 A. M.

Weather pleasant; on station off Morro.

THOS. C. WOOD.

4 to 8 A. M.

Left night station on blockade at 4.30 Spoke launch off Aguadores and learned she had been fired on during night, but no one hurt. Ran close in shore to eastward of Aguadores for reconnoisance. Heard rapid-fire gun at Aguadores. Came to position under west shore of bight.

GEORGE H. NORMAN, JR.

8 A. M. to noon.

Weather clear and pleasant. Heard three rifle shots between eight and nine o'clock, but saw no smoke. Observed no one on shore. At 10.30 saw a locomotive and one flat car going towards Aguadores from the eastward. Notified Flagship by wigwag. A few men seen on shore at ten o'clock.

J. T. EDSON.

Noon to 4 P. M.

Lying off Aguadores; weather clear, wind changeable with rain squalls. Received signals from Flagship, " Lebanon with stores for fleet will go between each two ships in succession "; also, " Look out for small steamer from Kingston, Jamaica; do not allow her inside line."

THOS. C. WOOD.

4 to 6 P. M.

At four o'clock signalled Flagship " In need of coal for steaming." Were then asked how much coal we had and answered " About six tons." We had informed Flagship at noon that we had but seven tons, and at noon for the ten days previous the ever-decreasing amounts of coal left us. Returned to station off Aguadores, two miles distant. At 4.50 Flagship signalled to come within hail. Were asked if we were correct in reporting but six tons. Answered " Yes." Were then ordered to proceed to Guantanamo at daylight tomorrow morning, and to return in time to reach fleet at dark, procuring as much coal as possible in the meantime. Returned toward Aguadores and, hearing some scattered rifle shots, by order of Executive Officer, lay to with one gun's crew at quarters from 5.35 till 6.50.

GEORGE H. NORMAN, JR.

6 to 8 P. M.

Squall to southeastward. Laid ship's head west by north. At seven o'clock left for station off Morro. Heard a few shots ashore on leaving.

J. T. EDSON.

8 P. M. to midnight.

On night blockading station off Morro.

THOS. C. WOOD.

GUANTANAMO, *Thursday*, June 30, 1898.

Midnight to 4 A. M.

On night blockading station; watch uneventful.

GEORGE H. NORMAN, JR.

4 to 8 A. M.

Left station off Morro at five o'clock, patent log reading 13.8, and set course E ¾ S. Weather clear and pleasant. The camp fires of the army are all close to the landing places. No smoke seen further in. At 6.30 set course east: patent log 30.7. At seven Revenue Cutter Manning passed to starboard heading west. At 7.15 passed *New York Herald* boat Mindora going west.

J. T. EDSON.

8 A. M. to noon.

Arrived at Guantanamo at nine o'clock. Reported to Marblehead, Commander McCalla, and received orders to coal from Justin. Went alongside and during watch took in twenty-six tons Pocahontas coal.

THOS. C. WOOD.

Noon to 4 P. M.

Coaling at Guantanamo. Transferred wherry to Justin. Knocked off coaling at 3.45, cast off, and ran down alongside Marblehead. Took in 61 tons of Pocahontas coal all told.

GEORGE H. NORMAN, JR.

4 to 6 P. M.

Left harbor of Guantanamo at 4.25 P. M.; patent log 53; course west by south. Changed course to west at 4.45. Changed to W ¾ N at six o'clock; patent log 72.2. The iron pier at Altares abeam at 6.25, distant four miles.

J. T. EDSON.

6 to 8 P. M.

Arrived at fleet off Santiago at 7.30. Reported arrival and coal received to Flagship. Received orders to take up night position off Morro. Stood in under Morro at eight o'clock.

THOS. C. WOOD.

8 P. M. to midnight.

On night station about south south east from Morro, distant one-half mile. Uneventful watch.

GEORGE H. NORMAN, JR.

SANTIAGO BLOCKADE, *Friday,* July 1, 1898.

Midnight to 4 A. M.

Weather clear and pleasant; lying off Morro at usual night station.

J. T. EDSON.

4 to 8 A. M.

Moved to our day position off Aguadores at daybreak. Cleared ship for action and awaited orders from Flagship. Flagship moved into position to eastward of Aguadores accompanied by Suwanee. Made preparations for bombarding Aguadores.

THOS. C. WOOD.

8 A. M. to noon.

In Aguadores bight. Shortly before nine o'clock Flagship opened fire on fortifications and western gulch and directed us to begin firing. Bombarded with rapid-fire for a few minutes and then more slowly till eleven o'clock.

GEORGE H. NORMAN, JR.

Noon to 4 P. M.

Shore batteries of Aguadores bombarded by New York, Suwanee, Newark and Gloucester. About one regiment of troops ashore approached within rifle range of the shore batteries, and stopped there without action. The batteries were thoroughly shelled, silenced, and the flag was dismounted.

J. T. EDSON.

4 to 6 P. M.

Lying off shore to eastward of Aguadores. Observed troops of our army, say about 500 men, on railroad line. These moved to eastward at five o'clock and at 5.15 were taken aboard train arriving from eastward. At 5.30 train disappeared down the line.

THOS. C. WOOD.

6 to 8 P. M.

Took night position in blockade after receiving signal from Flagship to relieve Newark of squadron's mail before five o'clock in the morning.

GEORGE H. NORMAN, JR.

8 P. M. to midnight.

Lying off Morro at usual night station. Morro bearing northwest about half a mile distant.

J. T. EDSON.

SANTIAGO BLOCKADE, *Saturday,* July 2, 1898.

Midnight to 4 A. M.

Occupying our usual night position off Morro to southward and eastward. Weather pleasant; bright moonlight; swell from southeast.

THOS. C. WOOD.

4 to 8 A. M.

At daylight, about 4.15, left night position in search of Newark, and were intercepted by Flagship and given our station in the bombardment to follow. Got squadron's mail and some packages, and Assistant Engineer Procter and his baggage from the Newark. Took station assigned us off Aguadores and cleared ship for action. On signal from New York, bombardment began about 6.30 and continued for an hour. Were then called to New York, held up on the way by the Newark, and again proceeded on further call from New York.

GEORGE H. NORMAN, JR.

8 A. M. to noon.

At eight o'clock we lay alongside the Flagship; Lieutenant Staunton came aboard, and course was laid for Siboney. At Siboney Lieutenant Staunton went ashore in our wherry and two warrant officers of the New York went ashore in 1st whaleboat. Weather clear and pleasant. Returned to New York at noon; Lieutenant Staunton and officers went onboard. Wherry slightly damaged in hoisting.

J. T. EDSON.

Noon to 4 P. M.

Delivered mail to following vessels of fleet; Brooklyn, Massachusetts, Oregon, Indiana, Iowa and Texas. Disappointed in not obtaining supplies from Celtic as this vessel sailed to eastward about 3.30. Weather pleasant. Occupied position off Aguadores.

THOS. C. WOOD.

4 to 6 P. M.

At 4.15 were called to Flagship and took Captain Chadwick to Siboney, arriving there at 5.05. Lay off and on till his return to the ship at 6.10.

GEORGE H. NORMAN, JR.

6 to 8 P. M.

Returned from Siboney with Captain Chadwick; took night blockading station.

A. M. PROCTER.

8 P. M. to midnight.

Off Morro; Castle bearing north north west, distant one-half mile.

J. T. EDSON.

SANTIAGO BLOCKADE, *Sunday*, July 3, 1898.

Midnight to 4 A. M.

Bright moonlight: occupying night position, Morro bearing N.
N W to N W, one-half mile distant.

THOS. C. WOOD.

4 to 8 A. M.

Left night station towards daylight, moving slowly to eastward.
At 5.30 were wigwagged by Flagship but given no message.*

GEORGE H. NORMAN, JR.

8 A. M. to 6 P. M.

Clear and warm; at 9.30 went to quarters for inspection. Captain
began inspection of ship and crew. At 9.43, Mr. Procter having
the deck, Spanish fleet came out of harbor, Gloucester being then
about 3000 yards southeast of Morro. Went to general quarters
and Executive took the deck. Opened fire at 3500 yards range
from after guns and forward starboard guns as they were brought
to bear. Started fire-room blowers and turned to starboard, decreasing range to 3000 yards. Four Spanish cruisers came out in
column and stood to westward. Slowed down to wait for torpedo-
boat destroyers, at the same time keeping up fire on cruisers from
port battery. Iowa, Indiana, Oregon, Texas, and Brooklyn engaged Cristobol Colon, Oquendo, Maria Teresa, and Vizcaya, all
standing to westward. Forts on shore kept up slow fire during
action. When larger vessels were well clear and the rear one
about 1500 yards to westward of Morro, destroyers Pluton and
Furor came out and followed in their wake. Opened rapid-fire on
them from starboard battery at 2500 yards range and ran engines
at full speed, heading course about W N W. Indiana signalled†

---

* The Gloucester was called with the intention of sending her to Siboney, but
the Admiral decided to go there himself in his Flagship. Had the Gloucester
been sent she would have missed the fight.—H. P. H.

† It was difficult to make out the flags through the smoke of battle, but the
Chief Quartermaster finally reported them. I looked in the signal book for
the corresponding signal and read it aloud to the Captain, "Gunboats will advance." To us it was a most important order, for we were about to enter the
zone of fire of the Indiana and the signal meant that Captain Taylor had recognized our position, and that we should, therefore, not be in danger from his fire.
It seems almost incredible that the flags actually displayed by the Indiana
had no reference at all to the Gloucester! Captain Taylor was signalling to
the fleet that the enemy's torpedo-boats had come out! The signal as we in-
terpreted it was very comforting but it had no effect upon our movements.
Wainwright had already directed me to close in on the enemy and to disregard
the fire from the Indiana as he later told me to disregard the fire from the Socapa battery.—H. P. H.

" gunboats will advance." After this the fight between Gloucester and apparently uninjured destroyers was a thing apart from the battle in which larger ships were engaged. Used starboard forward guns on leading vessel (Pluton) and starboard after and stern guns on Furor. Gradually increased speed to over 17 knots and slowly overhauled destroyers. After passing 2500 yards range, heard reports of automatic gun onboard Furor. Zigzag line of splashes from this gun approached ship to within 30 yards then reports ceased. At 10.15 opened fire with Colt automatic guns on destroyers, at 1200 yards range (Assistant Paymaster Alex. Brown's division). Pluton slowed down and showed signs of distress about this time. Concentrated our fire on Furor. Range was decreased to 600 yards when Pluton ran on the rocks about four miles west of Morro and blew up. Furor soon began to describe circles with starboard helm and ceased firing, being disabled. One of her crew waved a white towel and we stopped firing. Lieutenants T. C. Wood and G. H. Norman and Assistant Engineer A. M. Procter were sent to rescue the crews and save the prizes if possible. As Furor was on fire and burning rapidly, they took off the living and rescued all they could find in the water and on the beach. Pluton was among the rocks in the surf and could not be boarded. Meanwhile several explosions took place on Furor and about 11.30 she threw her bow in the air and turning to port slowly sank in deep water. The following were rescued from the destroyers, and are believed to be the only survivors: Furor, Commander Carlier, Lieutenant Arderius, and 17 petty officers and men; Pluton, Commander Vasquez, Lieutenant Boado, and 24 petty officers and men. While one of our boats was still ashore, seeing heavy clouds of smoke behind the next point, the ship was moved in that direction, all hands being at quarters ready for action. On rounding the point we found two men-of-war on the beach and burning fiercely aft, the majority of the crews crowded on the forcastles. Our boats under Lieutenant Norman and Ensign Edson put off to the nearer ship (Infanta Maria Teresa, Flag) and secured all on board by landing them on the beach through the surf. Lieutenant Norman formally received the surrender of the Spanish Commander-in-Chief and all his officers and men present, and as soon as all hands were transferred to the shore brought on board the higher officers, including the Admiral. Lieutenant Wood meanwhile rescued the survivors of Oquendo, the second of the burning vessels. Lieutenant Norman was sent to land with a small force to establish a camp on shore and hoist United States Colors. He took all rations that could be spared. By order of Captain Evans the Admiral and his staff were transferred to Iowa. All other unwounded prisoners were sent to Indiana and the twenty-two wounded were taken to Siboney and put on Olivette. One prisoner died on board. This vessel received no injury nor were any officers or men injured except Ensign Edson who had a rib broken by the recoil of a gun he was firing. New York passed us going west during the action and cheered Gloucester. Indiana signalled Gloucester after action, " Congratulations." Expended in Ordnance Department ammunition as follows: 6-pdr. Hotchkiss, 589 rounds; 3-pdr. Hotchkiss, 780 rounds; 6 mm. rifle (Colt automatic), 3400 rounds.

HARRY P. HUSE, Lieutenant and Executive Officer.

6 to 8 P. M.

Returning from Siboney. At 9.15 called " All hands bury the dead," and consigned dead Spanish prisoner to his grave.

THOS. C. WOOD.

8 P. M. to midnight.

Lying off Morro; weather pleasant. Bright moonlight.

A. M. PROCTER.

FUROR SINKING, 11.30 A. M., JULY 3, 1898.
From a photograph taken at the time onboard the Gloucester.

SANTIAGO BLOCKADE, *Monday*, July 4, 1898.
Midnight to 4 A. M.

Lying off and on Morro; weather pleasant, bright moonlight.

THOS. C. WOOD.

4 to 8 A. M.

Clear and pleasant. Lieutenant Staunton came alongside in torpedo boat Ericsson and hailed the ship. Congratulated the ship upon conduct in action and gave news of surrender of Colon. Flagship came in from westward about six o'clock. Commanding Officer reported aboard Flagship in obedience to signal. Proceeded to Siboney, with Harvard and took on board Lieutenant Norman and twelve men of his boat's crew.

A. M. PROCTER.

8 A. M. to noon.

Went on board Harvard with instructions to get list of prisoners and wounded on board that vessel. Returned at 12.20 with report of Captain Goodrich addressed to Admiral Sampson.

J. T. EDSON.

Noon to 4 P. M.

Lying off Harvard waiting for return of Mr. Edson with dispatches; at 12.20 Mr. Edson returned on board. Shaped course for Siboney. On arrival there at two o'clock sent Mr. Procter ashore with telegrams. On his return on board at 3.20 shaped course to westward to report to Flagship. Wind fresh from eastward; swell heavy.

THOS. C. WOOD.

4 to 8 P. M.

Sent Mr. Procter with dispatches to New York and on his return at six o'clock took telegrams he brought to Siboney sending them in by Assistant Paymaster Brown in second whaler. Whaler capsized in surf, but dispatches were saved and sent. Boat and crew returned to ship not seriously injured. Shaped our course for Flagship at eight.

GEORGE H. NORMAN, JR.

8 P. M. to midnight.

Returned to Flagship, delivered dispatches, and took night station. At 11.40 a shot was fired from Massachusetts. Went to general quarters. Massachusetts and Texas continued firing at entrance to harbor using search-lights. Massachusetts signalled to Flagship "I think that the enemy is trying to sink a vessel broadside on in the entrance." Flagship made circuit of fleet and returned to station.

A. M. PROCTER.

SANTIAGO BLOCKADE, *Tuesday*, July 5, 1898.

Midnight to 4 A. M.

Massachusetts and Texas firing occasionally at the entrance of the harbor. Piped down at 12.20; the remainder of the watch quiet and pleasant.

J. T. EDSON.

4 to 8 A. M.

Lying off Morro; night position.

THOS. C. WOOD.

8 A. M. to noon.

Sent by Flagship to Siboney for telegrams. At 9.05 sent first whaler ashore with Paymaster Brown. At 10.05 first whaler returned; stood away with telegrams for Flagship.

GEORGE H. NORMAN, JR.

Noon to 4 P. M.

Provisioning ship from Celtic. Received stores on board in Paymaster's department.

A. M. PROCTER.

4 to 6 P. M.

Clear and pleasant. Passing in view of the wrecks of the Oquendo, Maria Teresa and Pluton, heading east. Took station off Morro at five o'clock. As we passed Morro we could see the Mercedes ashore on eastern bank of entrance.

J. T. EDSON.

6 to 8 P. M.

Off Morro in day station. At dusk moved in to one mile.

THOS. C. WOOD.

8 P. M. to midnight

Moved into night blockading station at dark.

GEORGE H. NORMAN, JR.

SANTIAGO BLOCKADE. *Wednesday*, July 6, 1898.

Midnight to 4 A. M.

Weather pleasant; on station off Morro Castle.

A. M. PROCTER.

4 to 8 A. M.

Weather pleasant. Lying off Morro Castle. A commodore's pennant seen flying from Oregon at daylight. At eight o'clock Oregon was signalling New York.

J. T. EDSON.

8 A. M. to noon.

Occupying our usual day station. Scrubbing hammocks and clothes and engaged in cleaning ship.

THOS. C. WOOD.

Noon to 4 P. M.

Weather pleasant; on day station.

GEORGE H. NORMAN, JR.

4 to 6 P. M.

On station off Aguadores. Small tug bearing flag of truce came out of harbor at six o'clock.

A. M. PROCTER.

6 to 8 P. M.

Occupying station close under Morro Castle. Tug boat bearing flag of truce communicated with Suwanee and returned inside of harbor.

J. T. EDSON.

8 P. M. to midnight.

Lying close in under Morro watching entrance to Santiago harbor. Moon rose at 9.20. Then stood out to one mile and returned to usual night station. At eleven o'clock received signal from flagship New York to go alongside her at 7 A. M. prepared to receive Board of Survey and take it to wreck of Christobal Colon.

THOS. C. WOOD.

AT SEA, *Thursday*, July 7, 1898.

Midnight to 4 A. M.

About one o'clock launch of Flagship in charge of cadet, whose answer to boat hail was " New York," left written orders for Mr. McElroy to act on Board of Survey of wrecked Spanish ships and to go on board Suwanee at eight o'clock. Lying on Flagship's starboard quarter at four o'clock.

GEORGE H. NORMAN, JR.

4 to 8 A. M.

Started around fleet to pick up officers for Board of Survey on wrecked Spanish ships.

A. M. PROCTER.

8 A. M. to noon.

Lieutenant Nicholson came aboard from the Oregon, Lieutenant Haeseler from the Texas. Set course to westward. Board of Survey went aboard the Maria Teresa. At noon we were passing the Vizcaya lying on beach near Acerradero seven miles west of the Maria Teresa.

J. T. EDSON.

Noon to 4 P. M.

Standing to westward looking for wreck of Spanish cruiser Cristobal Colon. Board of Survey on board. Found cruiser on beach about 46 miles to westward of Morro. At 2.45 P. M. stood close in, and at 3.10 lowered boats with officers of Board who went in and made inspection of wreck.

THOS. C. WOOD.

4 to 6 P. M.

Standing off and on near wrecked Cristobal Colon.

GEORGE H. NORMAN, JR.

6 to 8 P. M.

Lying off wreck of Cristobal Colon; started ahead on course E by S at 7.23, patent log reading 35.8.

A. M. PROCTER.

8 P. M. to midnight.

Course E by S, making 60 revolutions of propeller. A white light seen from time to time on port bow slowly drawing aft. Weather clear; light air from the northward.

J. T. EDSON.

GUANTANAMO, *Friday*, July 8, 1898.

Midnight to 4 A. M.

Weather pleasant; swell heavy from southward. Ship standing off shore on course E by S; 60 revolutions of propeller. At two o'clock put her about, patent log 78.5, and stood inshore on course N W ¾ W. At 3.20 sighted a vessel two points on the port bow; exchanged private night signals, and soon made her out to be the Suwanee. Changed course six points to westward to clear her, giving two blasts of steam whistle to show our starboard helm. Resumed course N W ¾ W at 3.30.

THOS. C. WOOD.

4 to 8 A. M.

Standing in at slow speed. At five stood down to eastward toward Acerradero. At six sent Board to view wreck of Vizcaya and lay to awaiting their return.

GEORGE H. NORMAN, JR.

8 A. M. to noon.

Lying off wreck of Spanish ship Vizcaya. Headed back to fleet for orders, leaving Board of which Lieutenant-Commander Raymond P. Rodgers is senior member at scene of wreck. Commanding Officer reported on board U. S. S. Newark. Got under way for Guantanamo at 11.45 A. M.

A. M. PROCTER.

Noon to 4 P. M.

Weather pleasant; strong breeze from east. Came to anchor at Guantanamo at three o'clock. Relieved Mr. Wood at three.*

J. T. EDSON.

4 to 6 P. M.

Received permission to go alongside Celtic for supplies, and moored to her port side. At four o'clock began taking ice on board.

THOS. C. WOOD.

6 to 8 P. M.

Cast off from Celtic and laid our ship alongside schooner Wm. B. Palmer for coal by order of Admiral.

GEORGE H. NORMAN, JR.

8 P. M. to midnight.

Lying alongside coal schooner Wm. B. Palmer.

A. M. PROCTER.

---

*Mr. Edson had been on the sick-list with a broken rib (see log of July 3).

5

GUANTANAMO, *Saturday*, July 9, 1898.

Midnight to 4 A. M.

Alongside coal schooner Wm. B. Palmer in harbor of Guantanamo; weather pleasant.

THOS. C. WOOD.

4 to 8 A. M.

Prepared to begin coaling at daybreak. At five o'clock Captain McCalla ordered us away from the Wm. B. Palmer to the Mary E. Palmer. Began coaling about seven.

GEORGE H. NORMAN, JR.

8 A. M. to noon.

Coaling ship from schooner Mary E. Palmer.

A. M. PROCTER.

Noon to 4 P. M.

Coaling ship from Mary E. Palmer. Received stores and provisions as follows: 750 lbs. soap from Harvard, 20 cases 6-pdr. ammunition from Iowa, and 25 cases 3-pdr. ammunition from Vulcan. Transferred John Williams, coal passer, to U. S. S. Harvard for passage home to hospital, Navy Yard, New York.

THOS. C. WOOD.

4 to 6 P. M.

German and Japanese naval attachés came onboard to visit wrecks of Spanish ships. Coaling ship till 5.30. Sent mail for Niagara and Osceola on board Flagship. Picked up Lieutenant Wood in wherry as we passed Flagship. Stood out of bay and shaped a course to westward.

GEORGE H. NORMAN, JR.

6 to 8 P. M.

Standing on course west. Passed U. S. Flagship Newark. Newark signalled to come within hail, then ordered us to return to Guantanamo. At 7.32 started on course E ½ N for Guantanamo; patent log 55.2.

A. M. PROCTER.

8 P. M. to midnight.

Returned to Guantanamo and reported to Flagship. Ordered to await arrival of Suwanee with officers of Board and then resume course to westward to give the German and Japanese Naval Attachés an opportunity to visit and inspect wrecks of Spanish cruisers. Left Guantanamo at 11.20.

THOS. C. WOOD.

At Sea, *Sunday*, July 10, 1898.

Midnight to 4 A. M.

Weather pleasant; running to the westward at 50 revolutions. At three o'clock spoke U. S. S. Osceola. At four, passing the transport fleet off Siboney.

GEORGE H. NORMAN, JR.

4 to 8 A. M.

Standing on course W ¾ N. At 5.10 started ahead full speed, Daiquiri bearing abeam, patent log 63.1. Stopped alongside Brooklyn and received whaleboat and crew from her that had been left on Vizcaya. Heading for Acerradero.

A. M. PROCTER.

8 A. M. to noon.

Reached Acerradero at eight o'clock; lowered first whaleboat and sent her with Dr. Bransford in charge to inspect the wreck of the late Spanish cruiser Vizcaya with German and Japanese Naval Attachés. They returned to the ship at 10.30. Stood to the westward and arrived off wrecks of Oquendo and Maria Teresa at 11.30. Crew engaged in cleaning ship.

THOS. C. WOOD.

Noon to 4 P. M.

Sent away first whaler with German and Japanese Attachés and Mr. Procter in charge of Lieutenant Norman to visit wrecks of Oquendo and Infanta Maria Teresa and to drag bottom about bow of Teresa. Ran back and viewed wreck of Pluton. Hove to off wreck of Oquendo and Maria Teresa. At 2.25 whaleboat returned. At three o'clock stood away for Siboney.

GEORGE H. NORMAN, JR.

4 to 6 P. M.

Standing on to Siboney. Arrived off Siboney and hailed Wilmington for mail. Were hailed by St. Paul with orders to take a dispatch ashore to Colonel Turner. German and Japanese Attachés went onboard Segurança. At 5.50 shaped course for Guantanamo; patent log 78.

A. M. PROCTER.

6 to 8 P. M.

Bound to Guantanamo. Met flagship New York at 7.30 bound westward. Reported to her and proceeded. Received permission to water at Daiquiri.

THOS. C. WOOD.

8 P. M. to midnight.

Weather misty; dark. Standing to eastward. About ten o'clock made out Guantanamo, ran slowly in, and anchored in twelve fathoms of water half a mile to southward of other shipping. Iowa only ship to announce herself in response to our signals. Lieutenant Norman in dinghy reported Gloucester's arrival to her and learnt that Commodore Watson was in port on U. S. S. Newark. Sought mail on Vesuvius and Marblehead. Returned to ship and started out again to report to Commodore and seek further for mail. Reported at 11.55. Inquired for mail of New Orleans and boarded her to correct misunderstanding as to hail. Returned to ship at 12.30; relieved at 12.50.

GEORGE H. NORMAN, JR.

Midnight to 4 A. M.   OFF DAIQUIRI, *Monday*, July 11, 1898.

At anchor at Guantanamo; misty weather.

A. M. PROCTER.

4 to 8 A. M.

Got underway at 5.15 from Guantanamo and stood to westward towards Daiquiri for water. Vesuvius accompanying us, bound westward. Steaming slowly. Arrived off Daiquiri at eight o'clock.

THOS. C. WOOD.

8 A. M. to noon.

Weather pleasant though threatening; approaching Daiquiri from south and eastward. Off pier where Vixen lay taking water at 8.30. Went in wherry to inquire about watering; was informed by Lieutenant Harlow, commanding, Lieutenant Sharp being on sick list, that Vixen would lie there twenty-four hours more or any way till night, and that no other ship was seeking water. Osceola was then lying at anchor near by. Got provisions from transport and started for Flagship off Aguadores at 10.30, Captain Wainwright going on board on arrival. Got mail from Brooklyn. Flagship, Brooklyn and Indiana engaged in bombarding Santiago. Started for Siboney with dispatches from Flagship at 12.05. Relieved at one o'clock.

GEORGE H. NORMAN, JR.

Noon to 4 P. M.

Delivered dispatches at Siboney and proceeded to Daiquiri. Lying off Daiquiri during remainder of watch head to sea.

A. M. PROCTER.

4 to 6 P. M.

Off Daiquiri waiting our turn at dock for water; heading south by east, head to swell, moving slowly off shore. Weather threatening. Vesuvius passed bound west. Signalled and hailed that she had mail. Sent dinghy.

THOS. C. WOOD.

6 to 8 P. M.

Standing off shore to eastward of Daiquiri at slow speed. Running lights set at 6.20 for the first time since leaving Key West. Weather rainy. At eight o'clock, heading in shore stern to seas.

GEORGE H. NORMAN, JR.

8 P. M. to midnight.

Standing in slowly towards shore on course N by W. At 8.50 headed about on course S by E. Heavy rain squalls during watch.

A. M. PROCTER.

DAIQUIRI, *Tuesday*, July 12, 1898.

Midnight to 4 A. M.

Weather squally with heavy rains. Standing off shore off Daiquiri; course S by E. Head to swell; steaming slowly; wind variable.

THOS. C. WOOD.

4 to 8 A. M.

Standing out off shore from Daiquiri at dead slow speed on compass course S by E. Changed course and speed to N by W and slow at five o'clock. Changed course and speed at 6.30 to N N W, and dead slow.

GEORGE H. NORMAN, JR.

8 A. M. to noon.

Standing in towards Daiquiri. Changed course to S S E at 8.45. Heavy rain squalls. Headed in on course N N W at 11.15. Clearing to eastward at end of watch. Daiquiri in sight.

A. M. PROCTER.

Noon to 4 P. M.

Standing in towards Daiquiri. Arrived off harbor at 3.50. Sent boat ashore to arrange for taking water.

THOS. C. WOOD.

4 to 6 P. M.

Ran into Daiquiri and laid vessel parallel with pier forty feet distant, heading out.

GEORGE H. NORMAN, JR.

6 to 8 P. M.

Moored alongside pier at Daiquiri taking aboard fresh water.

A. M. PROCTER.

8 P. M. to midnight.

Moored alongside pier at Daiquiri taking fresh water.

THOS. C. WOOD.

SANTIAGO BLOCKADE, *Wednesday*, July 13, 1898.

Midnight to 4 A. M.

Moored off Daiquiri pier taking water.

GEORGE H. NORMAN, JR.

4 to 8 A. M.

Clear and pleasant; light northeasterly airs; barometer steady. Finished taking aboard fresh water at 4.20 A. M.

A. M. PROCTER.

8 A. M. to noon.

At 8.20 called all hands to unmoor ship. Cast off lines to pier and got underway at 8.30, having taken aboard 9000 gallons fresh water. Stood down to Siboney and sent ashore dispatches. Left Siboney at eleven o'clock and stood down to Aguadores and reported to flagship New York. By captain's order ——— ———, coal passer, was released from confinement and made a prisoner at large for twenty-four hours; then to resume his duty, awaiting approval of Commander-in-Chief of the sentence of Court Martial.

THOS. C. WOOD.

Noon to 4 P. M.

Standing by Flagship. At 12.30 started for Siboney to get Lieutenant Nicholson from U. S. S. Indiana and bring him back to Flagship. Sent wherry for Lieutenant Nicholson. Received a box from St. Paul. Returned with Lieutenant Nicholson to Flagship. Captain accompanied by Chief Engineer went on board Flagship. Paymaster returned during watch from Flagship and reported later that he had brought with him one thousand dollars.

GEORGE H. NORMAN, JR.

4 to 6 P. M.

Standing by Flagship. At five o'clock Chief Engineer returned on board accompanied by Lieutenant-Commander Raymond P. Rodgers. Received from Indiana for transfer to Solace and thence to Vermont Joseph Heddinger, seaman, Frank Ruich, fireman 1st class, and John Ludwig, gunner's mate 1st class. Took package from Massachusetts for Captain Evans of Iowa. Proceeded to Siboney and there sent aboard St. Paul for stores for Guantanamo, but were told that there were none aboard.

A. M. PROCTER.

6 to 8 P. M.

Running to eastward bound to Guantanamo with dispatches from Commander-in-Chief. Lieutenant-Commander Rodgers of Iowa passenger.

THOS. C. WOOD.

8 P. M. to midnight.

Weather pleasant, starlight; running to eastward from Siboney to Guantanamo. At 9.30 put ship on course north and ran into bay. At 10.10 anchored with 30 fathoms of chain in 8 fathoms of water, Fort McCalla bearing southeast, distant one-half mile. Sent Lieutenant-Commander Rodgers to Iowa in dinghy accompanied by Paymaster who then went to the Marblehead, thence to telegraph station ashore, returning to ship at eleven o'clock.

GEORGE H. NORMAN, JR.

GUANTANAMO AND SANTIAGO BLOCKADE, *Thursday*, July 14, 1898.
Midnight to 4 A. M.

At anchor in Guantanamo harbor. Watch uneventful.

A. M. PROCTER.

4 to 8 A. M.

At anchor in Guantanamo harbor. At six o'clock sent armature of forward dynamo to Vulcan to be turned off. Distributed mail to vessels in harbor. Engaged in breaking out forward hold to clean and paint.

THOS. C. WOOD.

8 A. M. to noon.

Weather hot; calm. In answer to our inquiries, Iowa signalled, "Solace has gone to Hampton Roads." Iowa's launch left letter for Captain at 8.50. Ran up quarantine flag at 8.52. Asked permission to get underway at 9.05; granted at 9.20. Weighed anchor about 9.30. At 10.07 stood away on course W S W ¾ W, patent log reading 49.1. At 10.50 break from high land to low coast line was abeam, patent log reading 58.4. At 11.05 changed course to W ½ N. At 12.08 changed course to W N W ½ W.

GEORGE H. NORMAN, JR.

Noon to 4 P. M.

Stopped off Daiquiri to send mail to Manning then proceeded to Flagship. Captain went aboard Flagship. Standing by Flagship. Flagship signalled "the enemy has surrendered."

A. M. PROCTER.

4 to 6 P. M.

Strong breeze from east south east. Oregon and Massachusetts left the squadron at 5.30 headed out. Wind moderating. Lying within hail of Flagship.

J. T. EDSON.

6 to 8 P. M.

Lying near Flagship.

THOS. C. WOOD.

8 P. M. to midnight.

Weather pleasant; dark night. Lying near Flagship. At 8.50 were signalled to take position for night at harbor entrance; then to come within hail; then told to stand by. Stood by the rest of watch. At midnight were called within hail again.

GEORGE H. NORMAN, JR.

Midnight to 4 A. M.   GUANTANAMO. *Friday*. July 15. 1898.

Standing by Flagship. At 12.05 Flagship signalled " come within hail." and then made general signal " Send boat to Flagship." Boat returned from Flagship with mail and dispatches for Guantanamo. Set course E S E ½ E at 12.50, patent log 85.4. At 2.20 changed course to east, patent log 99.5. At 3.34 changed course to E N E ½ E, patent log 10.5.

A. M. PROCTER.

4 to 8 A. M.

At 4.30 increased speed to 100 revolutions. Dropped anchor in harbor of Guantanamo at 5.30 in 7 fathoms of water with 30 fathoms of chain. Received orders from Marblehead to fly quarantine flag. The other vessels flying this flag are St. Paul, Manning, and tug Dandy. Hailed by Oregon to keep strict quarantine.

J. T. EDSON.

8 A. M. to noon.

At anchor in Guantanamo harbor. Sent mail and dispatches to Marblehead to be transferred to St. Paul for United States. At 11.50 signalled for and received permission from flagship Newark to get underway.

THOS. C. WOOD.

Noon to 4 P. M.

Weighed anchor at 12.20; went within hail of Newark and were held there till 1.10. Stood out of bay. Wheel rope shackle opened and steam steering gear thrown out at 1.40. Kept ship underway, and at 2.40, steering gear having been repaired, stood away for Aguadores; course W ½ N, patent log reading 11.3; Guantanamo entrance bearing N by E, distant two miles. At three o'clock changed course to W ¾ N. Break from high to low coast line bore abeam at 3.09, patent log reading 17.6. Weather misty.

GEORGE H. NORMAN, JR.

4 to 6 P. M.

Daiquiri abeam at 4.30. Went alongside Flagship and Commanding Officer went aboard. Standing by Flagship remainder of watch.

A. M. PROCTER.

6 to 8 P. M.

Lying off Santiago within signalling distance of Flagship; weather pleasant; light rains from westward.

J. T. EDSON.

8 P. M. to midnight.

Weather pleasant; lying off Aguadores within signalling distance of Flagship.

THOS. C. WOOD.

SANTIAGO BLOCKADE, *Saturday,* July 16, 1898.

Midnight to 4 A. M.

Weather pleasant; light land breeze. Standing by flagship New York off Aguadores.

GEORGE H. NORMAN, JR.

4 to 8 A. M.

Pleasant weather. Flagship made general signal to Gloucester, " Board steam vessel," and wigwagged, " Board strange vessel to southward, she is German." Proceeded to obey signal.

A. M. PROCTER.

8 A. M. to noon.

At 8.10 Lieutenant Wood boarded the Senior lying near Oquendo and learned that she was engaged as wrecking boat. Returned to Flagship off Aguadores, and sent Lieutenant Wood onboard. Weather pleasant; light airs from eastward.

J. T. EDSON.

Noon to 4 P. M.

Lying off Aguadores near Flagship; engaged in carrying dispatches between Flagship and Army signal station two miles to eastward on shore. At two o'clock, Captain Chadwick, Chief of Staff, came aboard for passage to Siboney. Left Aguadores at 2.15 and arrived at Siboney at 2.40. At three o'clock landed Captain Chadwick and Lieutenant Wood.

THOS. C. WOOD.

4 to 6 P. M.

Standing out of Siboney at four o'clock with dispatches for Flagship at Aguadores; delivered dispatches at 4.40. Returned to Siboney, arriving at 5.40.

GEORGE H. NORMAN, JR.

6 to 8 P. M.

Lying off Siboney awaiting return of Chief of Staff. Sent boat in for dispatches.

A. M. PROCTER.

8 P. M. to midnight.

Lying off Siboney awaiting return of Captain Chadwick. Boat returned with Chief of Staff and Lieutenant Wood about 10.30. Stood west for Flagship and came to anchor off Aguadores at midnight. Hailed by Flagship and told to lay off harbor of Santiago and prevent any one passing in or out. Chief of Staff went aboard Flagship.

J. T. EDSON.

OFF SANTIAGO, *Sunday*, July 17. 1898.

Midnight to 4 A. M.

Got up anchor at 12.10 and stood to westward; on reaching Morro, lay close in watching entrance; castle bearing north. Remained there during watch.

THOS. C. WOOD.

4 to 8 A. M.

Under Morro. At 5.30 observed a small squad of men in both eastern and western batteries going from gun to gun. Weather clear, winds light.

GEORGE H. NORMAN, JR.

8 A. M. to noon.

Off Morro. At 8.30 Vixen came within hail and reported that she had been ordered to assist in removing mines. Signalled to Flagship, "Am I to have directions about removing mines at 9 A. M.?" Flagship steamed in to entrance of harbor.

A. M. PROCTER.

Noon to 4 P. M.

Left Flagship off Morro at 12.15 with telegram to be sent at Siboney. Mr. Wood took dispatch ashore at 1.40. Sent dispatch aboard Flagship at 2.20. Gunboat Alvarado with American flag came out of harbor and visited New York and Brooklyn, at 3.30 P. M. Hailed by Flagship and told to tell the Adria that she might follow State of Texas into Santiago and to leave Merrimac on the starboard hand. Weather pleasant.

J. T. EDSON.

4 to 6 P. M.

Lying off Morro near Flagship; weather pleasant; swell from southward and eastward.

THOS. C. WOOD.

6 to 8 P. M.

Lying off entrance of Santiago. At seven o'clock called within hail of Flagship and given telegrams to be sent from Siboney. Siboney bore E N E at eight o'clock.

GEORGE H. NORMAN, JR.

8 P. M. to midnight.

Sent dispatches ashore at Siboney. Returned to Flagship bearing dispatches for Admiral. Standing by Flagship.

A. M. PROCTER.

OFF SANTIAGO, *Monday*, July 18, 1898.

Midnight to 4 A. M.

At three o'clock were called within hail by Flagship and given dispatches to be sent from Siboney. Weather pleasant.

J. T. EDSON.

4 to 8 A. M.

Entered Siboney harbor; sent Lieutenant Wood ashore with dispatches to telegraph office. At 5.30 on his return left harbor and returned to Flagship off Morro.

THOS. C. WOOD.

8 A. M. to noon.

Weather pleasant. Off Santiago entrance, Morro distant two miles and Flagship one and a half mile to the northward. At eight o'clock Vixen, flying Admiral's flag, passed in to entrance. Swung ship from 8.15 to 8.45 for compass observations. At nine, went alongside Flagship and were hailed to send a boat. At 9.30 received dispatches and stood away for Siboney. At 10.30 sent Paymaster ashore with dispatches. Paymaster returned on board with telegrams for Flagship at 11.15. Approaching Flagship off Morro at noon.

GEORGE H. NORMAN, JR.

Noon to 4 P. M.

Sent dispatches aboard Flagship; boat returned with bag of army mail. Standing by Flagship to end of watch. Army transports began entering harbor at three o'clock. The following men-of-war arrived: Dupont, Potomack, Leyden, and Suwanee. Small American flag was hoisted on Morro about 1.30 P. M. Quarterly Board of Survey of which Lieutenant Wood is senior member condemned 50 lbs. of fresh beef as unfit for issue. The beef was thrown overboard.

A. M. PROCTER.

4 to 6 P. M.

Lying off Santiago within call of Flagship. Twelve transports entered harbor. Weather pleasant.

J. T. EDSON.

6 to 8 P. M.

Off Morro in attendance on Flagship: weather pleasant: heavy swell from southeast.

THOS. C. WOOD.

8 P. M. to midnight.

Weather pleasant: sky cloudless. Morro distant two and a half miles to northward. At about nine o'clock Flagship approached and lay on our port quarter: a little later steam launch coming from direction of Flagship passed close under our stern and though hailed four times by our after lookout would give no answer. Launch disappeared in direction of U. S. S. Brooklyn. At midnight Morro bore north by east, distant two miles. Flagship close aboard.

GEORGE H. NORMAN, JR.

GUANTANAMO, *Tuesday*, July 19, 1898.

Midnight to 4 A. M.

Lying off Morro within signalling distance of Flagship. Weather pleasant; freshening breeze.

A. M. PROCTER.

4 to 8 A. M.

Weather clear and pleasant. Lying south of Morro; Flagship and Brooklyn to the northwest; Iowa to the southeast; Vixen inshore. Vixen was called by Flagship at seven o'clock and afterwards steamed to the eastward.

J. T. EDSON.

8 A. M. to noon.

Off Santiago harbor in attendance on Flagship. At 8.30 Captain and Executive visited Flagship and returned at 9.30. Got underway for Guantanamo; took departure from Morro at 9.50 (Flagship time) Morro bearing abeam (N ¾ E), distant two miles, patent log 45; course E ¾ S. The Board of Survey (consisting of Lieutenant Wood, Dr. Bransford, and Assistant Engineer Procter) having inspected three pounds of flour belonging to No. 2 mess and fifty pounds fresh meat belonging to messes Nos. 2 and 3 found it all unfit for use and it was thrown overboard. By order of Commanding Officer the following punishments were ordered: ———— ————, ordinary seaman, and ———— ————, ordinary seaman, for failing to turn in scrubbed hammock, reduced to 2nd conduct class. At 10.32 railroad bridge at Siboney bore abeam (N ¾ E), distant three miles, patent log 54.2. At 10.52, west end of pier at Daiquiri bore abeam (N ¾ E), distant three miles, patent log 57.8. At 11.26 Baracos Point bore abeam (N ¾ E), distant one and a half mile, patent log 63.8; changed course to E ¼ N. At 11.15 passed a French man-of-war bound eastward.

THOS. C. WOOD.

Noon to 4 P. M.

Weather pleasant; bound for Guantanamo. Made following courses: To Rock Promontory E ¼ N; Rock Promontory abeam at 12.25, patent log 74.1. Break from high to low coast line abeam at 12.34, patent log reading 75. Leeward Point abeam at 1.17, patent log 84. Came to anchor at 1.45. Weighed anchor at two o'clock and anchored near Celtic at 2.20. Took stores for wardroom.

GEORGE H. NORMAN, JR.

4 to 6 P. M.

Anchored at Guantanamo.   Painting.

A. M. PROCTER.

6 to 8 P. M.

Got up anchor at seven o'clock and steamed alongside of schooner Frank A. Palmer and made fast. No air stirring.

J. T. EDSON.

8 P. M. to midnight.

Alongside schooner Frank A. Palmer; weather pleasant.

THOS. C. WOOD.

GUANTANAMO, *Wednesday*, July 20, 1898.

Midnight to 4 A. M.

Alongside schooner Frank A. Palmer; weather pleasant; watch uneventful.

GEORGE H. NORMAN, JR.

4 to 8 A. M.

Alongside schooner Frank A. Palmer. Began coaling at 5.30 A. M.

A. M. PROCTER.

8 A. M. to noon.

Coaling ship, taking in about seven tons an hour. Had received thirty-two tons by noon. Marblehead signalled us to send boat to Hawk for mail. Flagship New York entered harbor at nine o'clock. Oregon saluted with eleven guns. At noon she again saluted with thirteen guns.

J. T. EDSON.

Noon to 4 P. M.

Coaling ship from schooner Frank A. Palmer. Took in about twenty-eight tons during watch. Marblehead signalled that Hawk had mail for us; sent boat for it. Communicated with Vulcan regarding repairs to dynamo engine.

THOS. C. WOOD.

6 to 8 P. M.

Weather pleasant; alongside collier Frank A. Palmer. Knocked off coaling at 4.45, having taken in 61 $\frac{4.10}{25}$ tons. From 4 to 8 A. M. seven tons were taken in; from 8 to noon twenty-seven tons: from noon to 4 P. M. twenty-two tons; and from 4 to 4.45 P. M. five tons.

GEORGE H. NORMAN, JR.

8 P. M. to midnight.

Weather clear and pleasant, little or no air stirring. Lying alongside of schooner Frank A. Palmer.

J. T. EDSON.

GUANTANAMO, *Thursday*, July 21, 1898.
Midnight to 4 A. M.

Weather pleasant; alongside schooner Frank A. Palmer.

THOS. C. WOOD.

4 to 8 A. M.

Weather pleasant. Turned to coaling at 5.30. Collier ordered away by Captain McCalla at 6.30. Got permission to be laid alongside of collier again at seven o'clock.

GEORGE H. NORMAN, JR.

8 A. M. to noon.

Alongside schooner Frank A. Palmer coaling ship. Men from Vulcan working on board repairing dynamo engine. Overhauled tiller and steering gear aft.

A. M. PROCTER.

Noon to 4 P. M.

Weather clear and pleasant. At 2.30 cast off from schooner Frank A. Palmer and came to anchor near by. Hoisted boats and prepared for sea. At four o'clock lay on starboard side of Massachusetts steaming slowly out of harbor in company with Massachusetts and Puerto Rico expedition. The following men were transferred back to the Indiana: Joseph Heddinger (Sea), Frank Ruich (Fireman 1st class) John Ludwig (G. M. 1st Class).

J. T. EDSON.

4 to 6 P. M.

Weather pleasant. Were ordered back to Yankee to get mail, packages, and freight for Massachusetts. Lying near Yankee receiving mail, packages, and freight. Received on board for transfer to U. S. S. Massachusetts Robert Johnson, seaman.

GEORGE H. NORMAN, JR.

6 to 8 P. M.

Standing by Yankee. Received on board boxes of stores and machinery parts for convoy fleet.

A. M. PROCTER.

8 P. M. to midnight.

Steamed slowly out of harbor at 8.30 P. M. At 8.53 put out patent log, reading 84.8, and set course east. Overtook and spoke steamer Stillwater, a press boat, out from Guantanamo. At 10.15 changed course to E by N, patent log reading 0. Reduced speed at eleven o'clock from 95 revolutions to 80 revolutions. Ten or more ships' lights on port bow drawing slowly aft, the Stillwater being abeam.

J. T. EDSON.

AT SEA, *Friday*, July 22, 1898.

Midnight to 4 A. M.

Weather pleasant. At midnight center of convoy bore abeam distant two miles. Speed 80 revolutions. Reduced speed to 65 at 12.05. At three o'clock with center of convoy a little forward of beam reduced speed to 50 revolutions. At 3.30 with relative positions same, reduced speed to 45 revolutions. At four o'clock with relative positions still unchanged reduced speed to 40 revolutions. At three o'clock sighted a steamer on port bow; she crossed our bows and passed us to starboard.

GEORGE H. NORMAN, JR.

4 to 8 A. M.

Convoy fleet on port beam. At 4.55 started ahead at full speed to overhaul Massachusetts. Massachusetts made signal, "Come within hail." Hailed Massachusetts and were ordered to take position 800 yards astern.

A. M. PROCTER.

8 A. M. to noon.

At 9.30 fleet was signalled to stop. Ordered by Flagship to transfer to her boats the freight, mail and smaller packages held by us for the other vessels of the fleet. Also transferred Robert Johnson, seaman, to Massachusetts. Started ahead again at ten o'clock, and took station previously held by Columbia. About 10.30 were signalled to board stranger which was headed north. Mr. Edson went in second whaleboat with armed crew and reported her to be the brig Goldfinch of Liverpool bound for New York City with cargo of lignum vitae and sugar. Her log showed she had sailed from Aza, San Domingo, on July 17. Returned to station about 12.30 P. M.

J. T. EDSON.

Noon to 4 P. M.

Weather pleasant. At 12.30 were two miles astern of Flagship. At 12.45 were ordered to take our station and make report later. Took position at 560 yards distance, bearing 45° from Flagship's starboard quarter. At two o'clock, squadron stopped while Flagship's boat went to and returned from Yale. Columbia came up with squadron at two o'clock. At four o'clock we bore 45° from Flagship's starboard quarter, distant 600 yards.

GEORGE H. NORMAN, JR.

4 to 6 P. M.

Pleasant weather. Flagship made signal, "Tell those transports to sheer in behind you."

A. M. PROCTER.

6 to 8 P. M.

Weather clear and pleasant; on our station in the fleet; course E ¼ N. Signalled Specialist to sheer in and keep position astern of us.

J. T. EDSON.

8 P. M. to midnight.

Weather cloudy; wind increasing. Ran over on orders of Flagship and directed Specialist to keep in our wake; then returned to station. Specialist paid no heed to orders. Flagship signalled general order at 9.30 to proceed at seven knots, then for one hour proceeded at five knots; increased it by eleven o'clock to eight knots, and ran the next hour at eight and a half knots. Specialist at about 10.30 rounded sharply off to southward and stopped and with her the next convoy astern of her. Tried to signal this to Flagship but could not secure her attention. At twelve o'clock nearest transport in starboard column was four miles distant.

GEORGE H. NORMAN, JR.

At Sea, *Saturday*, July 23, 1898.

Midnight to 4 A. M.

Moderate to fresh breezes; barometer steady. Transports slowly dropping astern. Furled bridge awning. Massachusetts slowed down and changed course at 3.30. Slowed down to 60 revolutions and changed course E by S.

A. M. Procter.

4 to 8 A. M.

Slowed down to half speed at daylight for half an hour. At 3.50 Massachusetts signalled to make seven knots. At six o'clock Columbia steamed up through fleet and took station ahead of us. Massachusetts then signalled us to take position astern of her. Weather clear and pleasant; a slight shower only at seven o'clock. Chronometer wound.

J. T. Edson.

8 A. M. to noon.

Weather pleasant. Flagship asked how much coal we would require on arrival at San Juan; answered, " None immediately; thirty-five tons to fill up." Flagship asked " How much coal can you carry." Answered, " Ninety-two tons in the bunkers." At noon reported latitude, longitude, and coal.

George H. Norman, Jr.

Noon to 4 P. M.

At 12.20 Flagship signalled courses E S E ½ E. Flagship made wigwag signal, " What is your coal expenditure? " Answered, " General expenditure is eight and a half tons." Flagship signalled latitude 20° 03', longitude 73° 12' at noon.

A. M. Procter.

4 to 6 P. M.

At station in fleet in rear of Massachusetts. Fleet having stopped in previous watch, at 4.10 Flagship signalled " Go ahead." A slight rain squall struck the ship at seven o'clock.

J. T. Edson.

6 to 8 P. M.

Weather squally with rain. Occupying our station in fleet 800 yards astern of Massachusetts.

Thos. C. Wood.

8 P. M. to midnight.

Weather squally; starlight. Ran through three light rain squalls. Watch otherwise uneventful. At midnight in position; transports in starboard column hull down.   George H. Norman, Jr.

Sunday, July 24, 1898.

Midnight to 4 A. M.

Gentle to moderate breezes; barometer steady; weather pleasant. Flagship signalled, " Find transports in starboard column and make them close up." Hailed Specialist and following transports and ordered them to close up. Furled quarter deck awning and lashed boxes stowed on deck.

A. M. PROCTER.

4 to 8 A. M.

Weather squally; two light showers during watch. Occupying station in rear of Flagship. At daylight Massachusetts signalled, " Close," and slightly reduced her speed for half an hour. Columbia and Yale closed in nearer Flagship. At eight o'clock the transports are getting into better position.

J. T. EDSON.

8 A. M. to noon.

Weather pleasant; in squadron astern of Flagship. At nine o'clock signalled latitude and longitude to Flagship. At 10.10 proceeded on new course S by W ½ W. At 11.45 took position at head of starboard column in place of Columbia. Dixie left squadron and stood to eastward.

THOS. C. WOOD.

Noon to 4 P. M.

Weather pleasant. Our position in squadron is leading vessels in starboard column.

GEORGE H. NORMAN, JR.

4 to 6 P. M.

Pleasant weather. At 5.30 P. M. Flagship signalled course south. Also signalled, " Extinguish all lights to-night; Flagship will show truck and stern lights."

A. M. PROCTER.

6 to 8 P. M.

Clear and pleasant; course by Flagship at seven' o'clock was S ¼ E.

J. T. EDSON.

8 P. M. to midnight.

Weather pleasant: wind fresh from eastward. Occupying our station 600 yards on starboard quarter of flagship Massachusetts. Moon set at ten o'clock. At 8.30 to 9.30 passed Mono Island on starboard beam, distant six miles.

THOS. C. WOOD.

LANDING PARTY.

GUANICA, PUERTO RICO, *Monday*, July 25, 1898.

Midnight to 4 A. M.

Weather pleasant; starlight and squally. Leading starboard column of squadron, course S E 7½ E. At 2.40 changed course to east and ran one mile; at 2.50 changed course to E by N and ran 4.9 miles; at 3.40 changed course to east and ran 3.3 miles to four o'clock. At four o'clock none of starboard transports in sight.

GEORGE H. NORMAN, JR.

4 to 8 A. M.

At daylight land appeared off port beam. All transports accounted for. At 5.30 Flagship changed course to N N W and signalled to come within hail. Came alongside and received orders to proceed to Guanica harbor. Following wigwag messages received and sent: Massachusetts to Gloucester, "Do you see any signs of a fortification?" Answer, "No, see Spanish flag on warehouse."

A. M. PROCTER.

8 A. M. to noon.

Lying off entrance of Guanica harbor waiting for fleet to come up. Made signal to Massachusetts, "Shall I go in"; answered, "Yes, you can try it." At 9 A. M. entered harbor in advance of the fleet. No guns could be seen on either side of the entrance. A few people were moving about near the light house. As the town came into view a Spanish flag was seen flying from a high staff. One bow gun was fired into one of the bluffs as a signal; no attention was paid to this and we could see people leaving the town by different routes. Our next shot was sent over the flag staff at a high elevation. This firing was not returned and the Spanish flag was not hauled down. Lieutenant H. P. Huse and Lieutenant T. C. Wood went ashore with an armed boat's crew, lowered the Spanish flag and hoisted ours. The men aboard ship cheered to see our flag ashore. Almost immediately after this a rapid firing of rifles was heard and we became aware that our men had been attacked; many rifle bullets struck the water alongside us and went singing past. Lieutenant Huse hailed us, requesting us to fire over him. The after guns were elevated to 2000 yards and fired repeatedly by Ensign J. T. Edson and Assistant Surgeon J. F. Bransford in the direction indicated. We could hear the boat's crew ashore using their Colt gun and rifles, and also the fire of the Spanish. Lieutenant Huse signalled that 250 men were needed to hold the place. Another armed boat was sent in charge of Lieutenant G. H. Norman and Assistant Engineer A. M. Procter and by this time some boats of the Massachusetts had entered the harbor. Lieutenant Huse returned with landing party, having left Lieutenant Wood on shore with a squad of men and the Colt gun at request of General Gilmore, U. S. A. After our work was done and the troop ships were anchored in the harbor, General Miles came on board and complimented Captain Wainwright on the manner of his action during the forenoon. Expended in Ordnance Department ammunition as follows: 16 6-pdr.; 60 3-pdr.; 500 6-mm. for rifle; 600 6-mm. for Colt automatic.

J. T. EDSON.

Noon to 4 P. M.

At anchor in Guanica harbor; weather warm and pleasant; gentle airs from S S E and S E. Transports engaged in landing troops. Remainder of landing party returned on board under Lieutenant Wood with Colt gun.

GEORGE H. NORMAN, JR.

4 to 8 P. M.

Weather pleasant; gentle airs from south south east veering to east at five o'clock and east north east at seven. Barometer steady. Transports Lampasas, No. 23; Stillwater, No. 28; Nueces, No. 1; Comanche, No. 11; Rita, No. 5; City of Macon; Specialist, No. 4; and Unionist, No. 31, anchored in harbor. Large vessels remained outside. Specialist aground on western side of harbor.

A. M. PROCTER.

8 P. M. to midnight.

Weather fair; gentle air from east north east veering to north at nine o'clock. Using search-light on surrounding hills. Scattered shots heard at intervals.

J. T. EDSON.

GOING INTO GUANICA.

Copyright 1898, by R. N. Davis.

THE GLOUCESTER FIRING OVER THE HEADS OF THE LANDING PARTY AT GUANICA.

The lighter brought off by Mr. Procter is alongside.   The steam launch was sent in from the Massachusetts
under the command of Naval Cadet F. T. Evans and helped to tow the lighter.

GUANICA, PUERTO RICO, *Tuesday*, July 26, 1898.

Midnight to 4 A. M.

Lying at anchor in Guanica harbor, acting as guard ship. Played search-light on hills about harbor to assist army in detecting presence of enemy.

THOS. C. WOOD.

4 to 8 A. M.

Weather hot and pleasant; sent away Massachusetts steam launch with Mr. Miller in charge at six to assist in towing lighters. Transports, with boats from the Massachusetts and lighters secured by us, busily engaged in landing men and stores.

GEORGE H. NORMAN, JR.

8 A. M. to noon.

Weather pleasant. The two steam launches of the Massachusetts engaged in aiding the transports to unload. Lieutenant Norman and Mr. Procter engaged in making a survey of the harbor. By order of Captain Higginson signalled Yale to "await orders." Captain Higginson left the harbor and went aboard Yale.

J. T. EDSON.

Noon to 4 P. M.

Lying at our anchorage at Guanica. Sent Lieutenant Murdock * ashore to confer with the Army Headquarters. Massachusetts signalled, "Dixie coming in." Ordered one of the transports to change her position so as to give the Dixie a good anchorage with proper bearings for her guns.

J. T. EDSON.

4 to 8 P. M.

Slight squall of rain. Occasional rifle shots heard ashore. Captain Wainwright and Lieutenant Huse called on General Miles on board of the City of Macon.

J. T. EDSON.

8 P. M. to midnight.

Massachusetts signalled us, "Prepare to get underway." There was considerable firing on the part of the pickets, and we were informed at the army headquarters that an attack was looked for. Captain Wainwright signalled Massachusetts, "Attack on Army, had better remain for the present." Massachusetts answered, " Remain." Firing of pickets heard ashore. Search-light gotten ready and the watch on deck ready at the guns.

A. M. PROCTER.

---

* A volunteer officer belonging to the Dixie. On detached duty at this time with U. S. troops. In compliance with naval customs he came onboard and reported to Captain Wainwright, senior officer present.

GUANICA, PUERTO RICO, *Wednesday*, July 27, 1898.

Midnight to 4 A. M.

Weather pleasant. Occasional rifle shots heard ashore. Picket lines are on the eastern hills near by. A few rifle balls have passed close to us and were distinctly heard. The Massachusetts signalled, "Send out a steam launch." The launch returned with Naval cadet Evans in charge with a message for General Miles stating that two transports had arrived with quartermaster's stores but no troops. Mr. Evans was directed to convey the message to General Miles on board the City of Macon. The lights of two steamers were seen coming from the east. Flagship signalled, "Transports remain outside and show truck lights."

J. T. EDSON.

4 to 8 A. M.

At anchor in Guanica; daylight at 4.45. Occasional rifle shots heard ashore.

THOS. C. WOOD.

8 A. M. to noon.

Crew engaged in cleaning ship. Captain Wainwright went in wherry to Wasp, and Lieutenant Huse with gig for Captain Paget,* who returned with him about nine. Captain Wainwright returned. Captain Higginson came on board and a few moments later Lieutenants Ward and Welles came onboard from the Wasp. At 10.30 the officers from the Wasp left taking with them Lieutenants Huse and Wood. Captain Paget returned aboard transport.

GEORGE H. NORMAN, JR.

Noon to 4 P. M.

Weather pleasant. Massachusetts made signal, "Tell General Miles will send Dixie, Annapolis, and Wasp to Ponce this afternoon at two o'clock." Wasp left harbor at 1.55 for Ponce in company with Dixie and Annapolis. Made signal to Massachusetts, "Will come out tonight unless receive further orders." Captain Higginson and Captain Paget visited the ship at 3.50. Received orders to proceed to Ponce.

A. M. PROCTER.

4 to 6 P. M.

Got up anchor at 4.40 and stood out. Captain Paget came aboard just as we were leaving. Stood in towards Ponce at six o'clock.

J. T. EDSON.

---

*Of the Royal Navy; British naval attaché.

6 to 8 P. M.

Underway for Ponce; arrived at 7.15 and anchored in harbor. Found Dixie, Annapolis and Wasp at anchor.

THOS. C. WOOD.

8 P. M. to midnight.

Weather pleasant; moonlight. Got underway. All lights aboard were extinguished or screened, and in silence with the crew at quarters and with a local pilot on the bridge we moved into the inner harbor as near as possible to the lighters and anchored. Two armed boats' crews under the Executive Officer (Lieutenant Huse) and Lieutenant George H. Norman pulled in and secured nine large lighters for use of army in landing, and towed them alongside. Then at ten o'clock got up anchor and slowly drifted back to the outer harbor to our anchorage near the Dixie. All through the manoeuvre the crew on board were kept at the guns, but our movements were not detected by the enemy.

GEORGE H. NORMAN, JR.

PORT PONCE, PUERTO RICO, *Thursday*, July 28, 1898.
Midnight to 4 A. M.

Pleasant weather; at anchor in harbor of Ponce. Watch uneventful.

A. M. PROCTER.

4 to 8 A. M.

Fleet sighted to southwest at 5.30. Boat from Dixie went ashore with flag of truce at 5.30. The fleet consisting of Massachusetts, Cincinnati, and Annapolis and four transports anchored in harbor at 7.30. Sent lighters to transports. Lieutenant Wood with armed boat's crew went in command of tug Cynthea. Weather pleasant.

J. T. EDSON.

8 A. M. to noon.

Weather pleasant; boat ashore with Lieutenant Wood to secure coal and water. Captain Paget, R. N., British naval attaché, left the ship.

GEORGE H. NORMAN, JR.

Noon to 4 P. M.

Ship visited by Captain Higginson to tell us that we were to leave what coal we were not loading at the time in lighters for the Wasp.

GEORGE H. NORMAN, JR.

4 to 6 P. M.

Ship visited by Lieutenant Ward, commanding Wasp, to tell us Captain Higginson had ordered coal loaded on lighter by us to be transferred to Wasp.

GEORGE H. NORMAN, JR.

6 to 8 P. M.

Weather pleasant; coal and water alongside in lighters.

J. T. EDSON.

8 P. M. to midnight.

At anchor in harbor of Ponce taking on coal from lighters on port side and water from lighter on starboard side. Natives doing the work.

THOS. C. WOOD.

PORT PONCE, PUERTO RICO, *Friday,* July 29, 1898.
Midnight to 4 A. M.

Knocked off coaling at 1.30. Sent coal lighter ashore in tow of Annapolis steam launch. Weather pleasant.

GEORGE H. NORMAN, JR.

4 to 8 A. M.

Pleasant weather; raining towards end of watch. Sent steam launch in to tow troops at 5.30.

A. M. PROCTER.

8 A. M. to noon.

Rainy weather. Natives engaged in delivering coal. Crew washing clothing and gun covers. Sent coal lighter ashore in tow of steam launch.

J. T. EDSON.

Noon to 4 P. M.

Weather cloudy. Board of officers convened on board to examine into matter of prizes in harbor.

THOS. C. WOOD.

4 to 6 P. M.

Weather pleasant, threatening clouds to the south and eastward. Captain Wainwright away with gig and Lieutenants Huse and Wood, Chief Engineer McElroy, Messrs. Edson and Brown went ashore. Finished taking water from lighters at 5.45. Annapolis came to anchor at 5.30.

GEORGE H. NORMAN, JR.

6 to 8 P. M.

Weather rainy; watch uneventful.

A. M. PROCTER.

8 P. M. to midnight.

Weather clearing; no air stirring.

J. T. EDSON.

PORT PONCE, PUERTO RICO, *Saturday,* July 30, 1898.

Midnight to 4 A. M.

Weather clearing; wind light from northward and eastward. Swell setting into harbor from southward. Moon set about 2.30. Shifted water lighter astern to avoid damaging her and the ship's side. Secured boats for sea as ship was rolling to swell.

THOS. C. WOOD.

4 to 8 A. M.

Weather bright. Called all hands at five o'clock and began carrying out morning orders. Sent Steward and Apothecary ashore at six o'clock on shore duty for two hours; sending wherry for them at 7.55.

GEORGE H. NORMAN, JR.

8 A. M. to noon.

Weather pleasant; cleaning ship. Wherry returned with Steward and Apothecary. Sent Chief Boatswain's Mate ashore for lighters, but he returned, unable to have them towed off.

A. M. PROCTER.

Noon to 4 P. M.

Got up anchor at 1.30 and stood out of harbor. Dropped anchor at 2.30 off light house of Muertos Island. Sent pilots ashore to arrange about having light resumed. Obtained sand from beach. Weather pleasant.

J. T. EDSON.

4 to 6 P. M.

Pilot and cutter returned; sent cutter ashore with four rations for the men at lighthouse. On return of cutter got up anchor and adjusted compass. At 5.30 stood to westward towards Ponce harbor.

THOS. C. WOOD.

6 to 8 P. M.

Came to anchor in Ponce harbor at old anchorage at 6.30. Dinghy returned with officers who had been ashore. Columbia reported from the eastward standing towards harbor. Captain went on board Flagship at 7.30.

GEORGE H. NORMAN, JR.

8 P. M. to midnight.

Captain returned on board at 8.30 with orders to get underway and proceed to place where Columbia was aground. Lieutenant Potts of Massachusetts came aboard under orders from Captain Higginson. Arrived alongside Columbia, sent Lieutenant Potts aboard, and took six-inch line from Columbia's quarter to windlass. Parted six-inch line. Took eight-inch line to windlass and backed off. Shifted line to port quarter, but were unable to move Columbia. Let go line. Anchored until Lieutenant Potts returned aboard, then returned to report to Captain Higginson.

A. M. PROCTER.

PORT PONCE, PUERTO RICO, *Sunday*, July 31, 1898.

Midnight to 4 A. M.

Lying at our anchorage off Ponce; weather pleasant. Mr. Procter returned to the ship at 12.15 with steam launch and lighter. Sent the lighter to Columbia.

J. T. EDSON.

4 to 8 A. M.

Weather pleasant; wind northeast. At 5.30 transports Obdam and Grand Duchess got underway and stood out of the harbor to westward. Columbia still aground at entrance of harbor. Crew engaged in getting coal and ice aboard.

THOS. C. WOOD.

8 A. M. to noon.

Finished getting in coal and ice and in obedience to signal from Massachusetts began collecting lighters for Columbia. Lieutenant Norman took ashore to General Miles Flagship's request that he send a steamer in aid of Columbia and obtained City of Macon. We then steamed out to Flagship and dropped collected lighters on Columbia's port side. Found it not advisable to obey Flagship's order to take a line from Columbia's starboard quarter and lay off within signalling distance of Flagship for an hour, when we were again ordered to take a line from the Columbia quarter which proved inexpedient and was not done. Press tug boats in response to Flagship's request came out to the stranded Columbia and gave what aid they could.

GEORGE H. NORMAN, JR.

Noon to 4 P. M.

Lying off Flagship ready to render aid to stranded Columbia. Captain went on board Flagship at 1.30 P. M., and returned at 2.30 P. M. Ordered to take one of Columbia's lines but were informed by Columbia that all lines were in use. Anchored at 2.30 P. M.

A. M. PROCTER.

4 to 6 P. M.

Lying at anchor near Flagship. Massachusetts and Cincinnati have lines to the Columbia. Wasp lying at anchor near us. Weather pleasant.

J. T. EDSON.

6 to 8 P. M.

Called by Flagship to take line from Columbia; engaged in doing this when Columbia got afloat. Stood in to harbor and in passing Flagship (Massachusetts) were ordered to take lighters back to harbor. Anchored and took lighters astern and anchored them till morning. St. Louis arrived off harbor and signalled she had mail for fleet. Sent boat.

THOS. C. WOOD.

8 P. M. to midnight.

At 9.30 boat returned with mail; at about ten o'clock Captain Higginson came alongside in launch and ordered us to send boat for Captain Goodrich at 5.30 in the morning and to then proceed with the Wasp to the harbor of Arroyo.

GEORGE H. NORMAN, JR.

ARROYO, PUERTO RICO, *Monday*, August 1, 1898.

Midnight to 4 A. M.

Weather pleasant; watch uneventful.

A. M. PROCTER.

4 to 8 A. M.

Weather pleasant. At 4.45 sent gig to St. Louis; gig returned at 5.30 with Captain Goodrich. Captain Lloyd of the Arcadia also came on board. Got up anchor at 5.45 and stood out to southward. At 6.30 stood to the eastward with pilot on deck. Two small sails sighted at 7.45 close in shore.

J. T. EDSON.

8 A. M. to noon.

At 8.15 stood in for Jobos; sent Lieutenant Wood in ahead to sound. Found only 2.5 fathoms and, as ships could not enter, stood on to eastward to Arroyo and anchored off that place at nine o'clock in 3.5 fathoms of water. Sent in Lieutenant T. C. Wood and Dr. Bransford to demand surrender of town under flag of truce. On landing, these officers met the officers of the town in the Custom House and demanded and received the formal surrender of Arroyo with all Spanish property in the name of Captain Goodrich, U. S. Navy, representing the United States. The only officer refusing to surrender being the Spanish Captain of the Port, who subsequently surrendered and gave his parole. Dr. Bransford was then sent back to the ship with these officials; the United States flag was hoisted over the Custom House, and Lieutenant Wood and Chief Quartermaster Bechtold remained on shore in charge. At 10.45 cutter and first whaleboat came ashore in charge of Lieutenant Norman with Paymaster Brown and the Colt gun, this force being all armed. On landing, pickets were stationed at all important points in the town. Lieutenant Norman was put in charge of right wing, Paymaster Brown of left wing with Colt gun commanding road to Guayama, and Chief Yeoman Lacy made Sergeant of the Guard. Headquarters were established at the Custom House, and orders given to hold the place and maintain order. The boats were anchored off the beach with keepers in charge.

THOS. C. WOOD.

Noon to 4 P. M

Lying off Arroyo. Landing party under Lieutenant Wood on shore in possession of town. Lifted anchor and shifted berth in shore. A deputation consisting of the Alcalde, a Justice of the Peace, and the Padre came off to discuss terms of surrender. Captain of Port came off under flag of truce.

A. M. PROCTER.

4 to 6 P. M.

Lying off Arroyo; two lighters were sent off from shore and anchored near the ship.

A. M. PROCTER.

6 to 8 P. M.

At seven o'clock Lieutenants Wood and Norman with infantry company returned to the ship. Weather unsettled and squally.

J. T. EDSON.

8 P. M. to midnight.

Lying at anchor off Arroyo in three fathoms of water distant 1000 yards from shore. Used search-light occasionally along the shore; all quiet.

J. T. EDSON.

ARROYO, PUERTO RICO, *Tuesday*, August 2, 1898.

Midnight to 4 A. M.

Weather pleasant. Lying off Arroyo in three and a quarter fathoms of water. Using search-light on shore. At 2.30 shots were heard in town. Called men to guns and turned light on shore, but all seemed to be quiet. One lighter brought off during watch.

A. M. PROCTER.

4 to 8 A. M.

Called all hands at four o'clock; at five called away cutter and whaleboat and sent landing party ashore to occupy Arroyo under charge of Lieutenant Wood, Lieutenant G. H. Norman and Paymaster Alex. Brown.

THOS. C. WOOD.

8 A. M. to noon.

St. Louis, St. Paul, and Cincinnati came to anchor in harbor. Two transports seen in offing. Lieutenant Wood in charge of men ashore signalled for us to fire at base of hill to his left, range one and a half miles. Four shots were fired. Lighters filled with troops were sent in from St. Louis towed by small boats. Sent food to men ashore.

J. T. EDSON.

Noon to 4 P. M.

At anchor off Arroyo. St. Paul and Cincinnati came in and anchored. St. Louis began landing troops.

A. M. PROCTER.

4 to 6 P. M.

Landing party returned, having been relieved by United States troops. Got underway and stood out to St. Louis.

HARRY P. HUSE.

6 to 8 P. M.

Came in and anchored off Arroyo. General Brooke and Aide and Captain Chester left ship in Cincinnati's steam launch. Signalled as follows: Gloucester to Cincinnati, "Annapolis and Puritan at Fajado to-day."

A. M. PROCTER.

8 P. M. to midnight.

Squally weather. Got up anchor and drifted to westward. Lying now with ten fathoms of chain in three fathoms of water. Signalled Cincinnati, "We are now lying as far to westward as safety will permit." Transports left harbor at 10.30 heading to west. All quiet.

J. T. EDSON.

ARROYO, PUERTO RICO, *Wednesday*, August 3, 1898.

Midnight to 4 A. M.

Rain squalls with wind from eastward. Dragging to leeward at one o'clock; got up anchor, steamed to windward a quarter of a mile and anchored in 3.5 fathoms with 15 fathoms of chain on starboard anchor.

THOS. C. WOOD.

4 to 8 A. M.

Weather pleasant; sky cloudy. Called all hands at 5.30 and carried out morning orders. At six o'clock Stillwater came in and anchored on our starboard quarter. Inquired for General Brooke aboard of us and then started out to St. Louis. Ordered by signal to tow lighters from St. Paul. Weighed anchor but signal was annulled. Crew went to breakfast at eight.

GEORGE H. NORMAN, JR.

8 A. M. to noon.

Got underway at 8.45 and stood out to Cincinnati in obedience to signal. Went out to Mississippi with orders from General Brooke that she go to Ponce. Went alongside of St. Paul. General Haines came onboard.

A. M. PROCTER.

Noon to 4 P. M.

At 12.30 General Haines went ashore with Lieutenant Wood in second whaler. Got up anchor and conveyed message to St. Paul, and then spoke transports 5 and 8 and piloted them into harbor. Came to anchor in four fathoms of water with fifteen fathoms of chain. Cincinnati firing occasionally to the left of the army on shore.

J. T. EDSON.

4 to 6 P. M.

At 4.30 started out to St. Paul in offing for stores and to give a message from General Haines. Stores not ready and returned to harbor. Anchored at 5.50 in three fathoms.

THOS. C. WOOD.

6 to 8 P. M.

At anchor off Arroyo; weather pleasant; light airs. Watch uneventful.

GEORGE H. NORMAN, JR.

8 P. M. to midnight.

As before. Press boat hailed ship with message from Major Carson asking that search-light be turned on landing place. Signalled message to Cincinnati. Cincinnati kept search-light on landing place during whole of watch.

A. M. PROCTER.

ARROYO, PUERTO RICO, *Thursday*, August 4, 1898.

Midnight to 4 A. M.

Clear and pleasant; light airs from eastward. Lying at anchor off Arroyo.

J. T. EDSON.

4 to 8 A. M.

Weather pleasant; swell from southward; light airs from eastward. At 5.30 St. Louis got underway and stood to westward. Signals passed between St. Paul in the offing and the Cincinnati. At seven o'clock New York *Journal* tug Echo arrived in harbor from westward. On shore everything apparently quiet. At 7.30 transport Gussie passed out from anchorage towing empty lighters to St. Paul.

THOS. C. WOOD.

8 A. M. to noon.

Weather hot; light airs. At anchor off Arroyo. At 10.30 requested permission of Cincinnati to go to St. Paul for stores. Repeated message at her request three times and were granted permission to proceed. Were then called within hail and held half an hour. Then proceeded to St. Paul with one of Cincinnati's boats in tow. Were informed by Captain Sigsbee that he had been unable to break out stores for Gloucester which he had promised twenty-four hours before. Sent Mr. Sprague on board to spend night.

GEORGE H. NORMAN, JR.

Noon to 4 P. M.

Sent steward with whaleboat and cutter for stores. Cincinnati signalled, "Send pilot to transport." Sent pilot to transport Roumania. Stood in and anchored.

A. M. PROCTER.

4 to 6 P. M.

Pilot from Roumania brought on board. At 5.15 Captain Chester came on board. Got up anchor and stood out to St. Paul. Returned to Cincinnati and then came to anchor at our usual station off Arroyo.

J. T. EDSON.

6 to 8 P. M.

Returned to anchorage off Arroyo at 6.30; weather pleasant.

THOS. C. WOOD.

8 P. M. to midnight.

At anchor off Arroyo; watch uneventful.

GEORGE H. NORMAN, JR.

ARROYO, PUERTO RICO, *Friday*, August 5, 1898.

Midnight to 4 A. M.

Weather pleasant; passing showers. Watch uneventful.

A. M. PROCTER.

4 to 8 A. M.

Everything quiet ashore; weather pleasant. At daybreak it was noticed that the St. Paul had left the harbor. A steamer was sighted passing to westward. Scrubbed decks and boat gear. Sent caterers ashore at 5.30; they returned at 7.40.

J. T. EDSON.

8 A. M. to noon.

Weather pleasant; Mr. Sprague of the Century Company, visiting ship from the St. Paul. Views taken of crew in action, etc.

THOS. C. WOOD.

Noon to 4 P. M.

At anchor off Arroyo; weather pleasant; light breeze from eastward. Watch uneventful.

GEORGE H. NORMAN, JR.

4 to 6 P. M.

Got underway and steamed out to St. Paul and put Mr. Sprague aboard. Returned to anchorage.

A. M. PROCTER.

6 to 8 P. M.

At anchor off Arroyo. St. Paul got up anchor and stood to the westward. Yacht Kanapaha arrived from Ponce. Weather pleasant.

J. T. EDSON.

8 P. M. to midnight.

Weather pleasant. Cincinnati played her search-light on shore during the whole of watch after nine o'clock. Forward dynamo gave out, extinguishing all electric lights in the ship.

THOS. C. WOOD.

ARROYO, PUERTO RICO, *Saturday*, August 6, 1898.

Midnight to 4 A. M.

Weather pleasant; sky cloudy. Watch uneventful. Lying at anchor off Arroyo.

GEORGE H. NORMAN, JR.

4 to 8 A. M.

Heavy rain squall at five o'clock. Scrubbing deck with sand. Field day.

A. M. PROCTER.

8 A. M. to noon.

Weather pleasant. Scrubbing and cleaning ship. Tug Leyden came into harbor from eastward.

J. T. EDSON.

Noon to 4 P. M.

Weather pleasant; crew engaged in putting ship in order for Sunday inspection. Newspaper yacht Kanapaha came into harbor from the eastward. Dynamo put in order by Chipman and Bee.

THOS. C. WOOD.

4 to 6 P. M.

At anchor astern of Roumania in Arroyo Harbor; weather pleasant.

GEORGE H. NORMAN, JR.

6 to 8 P. M.

Got underway and shifted berth nearer shore.

A. M. PROCTER.

8 P. M. to midnight.

Slight rain squall at 11.15; the two newspaper boats left the harbor at nine o'clock.

J. T. EDSON.

Arroyo, Puerto Rico, *Sunday*, August 7, 1898.

Midnight to 4 A. M.

Showers first part of watch; pleasant latter part; wind light from northward and eastward. Barometer steady.

THOS. C. WOOD.

4 to 8 A. M.

Weather fine; light breezes from eastward. Called all hands at five o'clock. At 5.45 sent market boat ashore with mess caterers and wardroom steward. At 6.30 the Summers N. Smith came in from westward and anchored on our port quarter; at 7.30 she weighed anchor and passed out standing to the eastward. At 7.10 sent wherry ashore to bring off market party.

GEORGE H. NORMAN, JR.

8 A. M. to noon.

Inspection by the Captain at 9.30 followed by general muster. Received signal from Cincinnati, "I am going ashore at once, meet me there." Captain Wainwright left the ship, accompanied by Dr. Bransford. Captain returned after conferring with Captain Chester. Sent signal to Cincinnati, "Please send steam launch." Steam launch immediately came alongside and left again for shore. Lieutenant Wood went ashore in steam launch to relieve Lieutenant Hoogewerff as beach-master.

A. M. PROCTER.

Noon to 4 P. M.

Cincinnati left the harbor at 12.30 standing to the westward. The steam launch is engaged in towing lighters. Dr. Bransford went ashore in first whaler at four o'clock to report to Lieutenant Wood.

J. T. EDSON.

4 to 6 P. M.

At anchor off Arroyo; weather fine.

GEORGE H. NORMAN, JR.

6 to 8 P. M.

At anchor off Arroyo. Lieutenant Wood and Dr. Bransford returned aboard at 7.30.

A. M. PROCTER.

8 P. M. to midnight.

Sent Bond in first whaler after steam launch at nine o'clock; returned about ten. Steamer Gussie went close in shore at ten. A slight squall of rain at eleven.

J. T. EDSON.

ARROYO, PUERTO RICO, *Monday*, August 8, 1898.

Midnight to 4 A. M.

At anchor off Arroyo.

GEORGE H. NORMAN, JR.

4 to 6 A. M.

Weather fair. Lieutenant Wood went ashore in dinghy at six o'clock. Steam launch went alongside Seneca for coal and water and then ashore to tow lighters. Arranged with Seneca about receiving coal.

A. M. PROCTER.

8 A. M. to noon.

Weather pleasant. ———— ————, seaman, was given five midwatches for having clothes in lucky bag; ———— ———— (o. s.), was given five midwatches for striking a petty officer in the execution of his duty. Steam launch engaged in towing lighters. Got up anchor and piloted transport Massachusetts and transport No. 28 to anchorage. Sent boat to transport Massachusetts for dispatches. Lieutenant George M. Stoney, temporarily in command of Massachusetts, visited the ship.

J. T. EDSON.

Noon to 4 P. M.

Weather pleasant; fresh breeze from eastward. Shifted anchorage to berth just ahead of Seneca, then passed line to lighters alongside Seneca, one filled with fifteen tons of coal and the other empty.

GEORGE H. NORMAN, JR.

4 to 6 P. M.

Hove lighters up alongside with steam windlass; commenced to coal ship. Coal stowed by laborers from shore.

A. M. PROCTER.

6 to 8 P. M.

Coaling ship. Sent whaleboat in for Lieutenant Wood. Steam launch returned for the night and was secured astern.

J. T. EDSON.

8 P. M. to midnight.

Lying at anchor off Arroyo. Coaling from lighters; having emptied lighter, anchored one astern and sent the other ashore in tow of steam launch.

GEORGE H. NORMAN, JR.

ARROYO, PUERTO RICO, *Tuesday*, August 9, 1898.

Midnight to 4 A. M.

Weather unsettled; watch uneventful.

A. M. PROCTER.

4 to 8 A. M.

Sent whaleboat ashore with Mr. Wood at six o'clock; returned at 6.30. Steam launch left ship under Lieutenant Wood's orders at 6.30. Saw Gussie at daybreak coming in from the west. Weather pleasant.

J. T. EDSON.

8 A. M. to noon.

At anchor off Arroyo; strong breeze; heavy roll.

GEORGE H. NORMAN, JR.

Noon to 4 P. M.

Captain Pitcher, U. S. A., of the transport Massachusetts requested steam launch to be sent to save lighters; steam launch was sent and after returning was engaged in towing lighters to and from shore. Tug Carbonero came into harbor at 3.30 P. M.

A. M. PROCTER.

4 to 6 P. M.

Sent cutter ashore for Lieutenant Wood, the Paymaster, and Whitelock. Steamer engaged in towing.

J. T. EDSON.

6 to 8 P. M.

At anchor off Arroyo; weather pleasant; heavy roll. Sent steam launch to tow lighter with coal from transport Massachusetts.

GEORGE H. NORMAN, JR.

8 P. M. to midnight.

Steam launch returned after having anchored coal lighters; secured steam launch astern and hauled fires.

A. M. PROCTER.

ARROYO, PUERTO RICO, *Wednesday*, August 10, 1898.

Midnight to 4 A. M.

Weather pleasant; watch uneventful.

J. T. EDSON.

4 to 8 A. M.

At anchor off Arroyo. Called all hands at five o'clock. Sent launch away at 5.50 for coal-barge, and later for workmen from transport Massachusetts. Sent away first whaler with Mr. Wood, Steward, and caterers at six. Sent in whaler at 7.10; she returned with Steward and without caterers.

GEORGE H. NORMAN, JR.

8 A. M. to noon.

Coaling ship from lighters alongside. Had quarters for inspection followed by infantry drill. Steam launch engaged in towing lighters to and from shore.

A. M. PROCTER.

Noon to 4 P. M.

At three o'clock sighted U. S. S. Annapolis to the eastward; entered harbor at four. Weather pleasant. The ground swell which has been noticeably increased during the past few days is gradually abating; weather clear and pleasant.

J. T. EDSON.

4 to 6 P. M.

Weather pleasant. Annapolis came in at five; Captain Wainwright went on board.

GEORGE H. NORMAN, JR.

6 to 8 P. M.

Returned steam launch to Annapolis.

A. M. PROCTER.

8 P. M. to midnight.

Visiting party to Annapolis returned at 8.15. Weather pleasant. The lights of two transports seen in the offing to the southwest.

J. T. EDSON.

ARROYO, PUERTO RICO, *Thursday*, August 11, 1898.
Midnight to 8 A. M.

At anchor off Arroyo; a rain squall between 2.30 and 3.30.

GEORGE H. NORMAN, JR.

4 to 8 A. M.

Transports expected by army not in sight. *Herald* dispatch boat Summers R. Smith came in and reported that terms of peace proposed by United States had been accepted by Spain. Annapolis signalled, "On account of news of peace I will carry out my orders and go to Ponce; will leave steam launch."

A. M. PROCTER.

8 A. M. to noon.

Heavy swell continues; wherry beached and gunwale torn away, towed out by whaleboat. Annapolis left for Ponce taking her steam launch with her. Rain squall at ten o'clock.

J. T. EDSON.

Noon to 4 P. M.

Weather unsettled; swell heavy from the southeast. Got anchor (purchased ashore) on the port bow; shackled chain and put it in readiness for use if required. At three o'clock vessel came in to anchorage which proved to be army hospital ship Relief. Made our number, but she did not respond and we learned her name in response to hail. As Relief carried Geneva Cross Flag at fore and a pennant of Naval Officer Commanding at main, Lieutenant Wood was sent to ask explanation and request that pennant be lowered if not authorized. Found Major Torney in command who promptly acceded to request.

THOS. C. WOOD.

4 to 6 P. M.

Weather threatening; sky overcast. At anchor off Arroyo; watch uneventful.

A. M. PROCTER.

6 to 8 P. M.

As before; rain ended about 7.30.

A. M. PROCTER.

8 P. M. to midnight.

Occasional showers of rain.

J. T. EDSON.

ARROYO, PUERTO RICO, *Friday*, August 12, 1898.

Midnight to 4 A. M.

Weather pleasant; swell somewhat subsiding.

THOS. C. WOOD.

4 to 8 A. M.

Weather pleasant; newspaper tug Hercules came in at six o'clock from St. Thomas and anchored close inshore.

GEORGE H. NORMAN, JR.

8 A. M. to noon.

Quarters at 9.30 followed by infantry drill. The following men from the army who had been on board for steam launch work were sent ashore: Privates A. B. Haynes, Hampton Battery B., Pennsylvania Volunteers; Leslie Dana Rumbaldt, Battery A., 1st Missouri Volunteers; Arthur King, 27th Battery, Indiana Volunteers; H. C. Hill, Battery A., 1st Illinois Volunteers. Major Carby of the U. S. Engineers came aboard with request from General Brooke for grapnels. Major Torney, Surgeon U. S. Army, called on the Captain.

A. M. PROCTER.

Noon to 4 P. M.

Army officers went ashore at one o'clock and Major Torney, commanding Relief, returned to his ship at two o'clock. Relief got underway and stood out of harbor bound to Ponce. Men of both vessels exchanged cheers as she crossed our bows. At 3.30 Lieutenant Huse and Lieutenant Wood went ashore to make official visits on army officers. Ensign Edson went on sick list.

THOS. C. WOOD.

4 to 6 P. M.

Weather pleasant. Lieutenants Huse, Wood, and Norman, and Paymaster Brown came off in second whaler at 4.30.

GEORGE H. NORMAN, JR.

6 to 8 P. M.

Lieutenant Huse went ashore at seven o'clock to confer with General Brooke, U. S. A., and returned at eight.

A. M. PROCTER.

8 P. M. to midnight.

Weather squally with occasional light showers.

J. T. EDSON.

ARROYO, PUERTO RICO, *Saturday,* August 13, 1898.

Midnight to 4 A. M.

Watch uneventful.

GEORGE H. NORMAN, JR.

4 to 8 A. M.

Rain squall at 7.30. Steward went ashore at seven o'clock. Field day.

A. M. PROCTER.

8 A. M. to noon.

Finished cleaning ship; weather overcast; light rain squall.

THOS. C. WOOD.

Noon to 4 P. M.

Weather pleasant. At two o'clock made out distinguishing pennant of U. S. S. Frolic in the offing. She anchored, distant five miles; ran out to her and received and sent away mail at four o'clock.

GEORGE H. NORMAN, JR.

4 to 6 P. M.

Returned and anchored off Arroyo. Frolic got underway and stood to eastward.

A. M. PROCTER.

6 to 8 P. M.

Weather pleasant; wind moderating.

THOS. C. WOOD.

8 P. M. to midnight.

Watch uneventful; received dispatch for Captain Wainwright about nine o'clock, signed " Rodgers."

GEORGE H. NORMAN, JR.

ARROYO, PUERTO RICO, *Sunday*, August 14. 1898.

Midnight to 4 A. M.

Weather pleasant. —— ——, ordinary seaman, and —— ——, seaman, stood mid-watch for punishment.

A. M. PROCTER.

4 to 8 A. M.

Weather pleasant; swell moderate from east south east. Engaged in preparing ship for inspection.

THOS. C. WOOD.

8 A. M. to noon.

Weather pleasant. Stillwater came in from westward at nine o'clock and anchored. Annapolis sighted on southeastern horizon at 11.30.

GEORGE H. NORMAN, JR.

Noon to 4 P. M.

Annapolis came in and anchored on port beam. Sent boat to Annapolis for mail. Commanding Officer reported onboard Annapolis in obedience to signal.

A. M. PROCTER.

4 to 6 P. M.

Weather pleasant; moderate swell from southeast.

THOS. C. WOOD.

6 to 8 P. M.

Weather pleasant; watch uneventful.

GEORGE II. NORMAN, JR.

8 P. M. to midnight.

As before; watch uneventful.

A. M. PROCTER.

ARROYO, PUERTO RICO, *Monday* August 15, 1898.

Midnight to 4 A. M.

Weather pleasant; starlight; wind light from eastward; swell moderate.

THOS. C. WOOD.

4 to 8 A. M.

Weather pleasant; watch uneventful.

GEORGE H. NORMAN, JR.

8 A. M. to noon.

Lieutenant Wood went aboard Annapolis and then ashore with Captain Hunker. Lieutenant Wood returned on board at eleven o'clock. Got underway for Ponce taking mail. Captain Hunt, U. S. A., and Mr. McCormick onboard as passengers for Ponce.

A. M. PROCTER.

Noon to 4 P. M.

Underway under one boiler bound to Ponce. At three o'clock passed a vessel on our port beam standing to the eastward which appeared to be an English war-ship. Reached Ponce at 3.55 and, securing permission, anchored in eight fathoms of water.

THOS. C. WOOD.

4 to 6 P. M.

At anchor in Port Ponce. Lieutenant Huse representing the Captain called on Captain Frederick Rodgers, commanding the Puritan, senior officer present. Captain Hunt went onboard the Columbia to take passage for home. Mr. McCormick went on shore.

HARRY P. HUSE.

6 to 8 P. M.

Weather pleasant; watch uneventful.

THOS. C. WOOD.

8 P. M. to midnight.

At anchor in Ponce harbor; weather pleasant, swell from southward and eastward.

THOS. C. WOOD.

PONCE, PUERTO RICO, *Tuesday*, August 16, 1898.

Midnight to 4 A. M.

Weather pleasant; watch uneventful.

GEORGE H. NORMAN, JR.

4 to 8 A. M.

Sent cutter in tow of Terror's steam launch to bring off laborers and lighters. Paymaster Brown went onshore to receive laborers and lighters and returned with same. Got underway and went alongside collier Saturn. Columbia left for Key West.

A. M. PROCTER.

8 A. M. to noon.

Coaling from Saturn. Finished at 11.30, having taken on board twenty-four tons. At 10.50 Flagship signalled for Commanding Officer to come aboard, and as Commanding Officer was indisposed, Lieutenant Huse reported and returned with orders to get underway at once for San Juan. Made preparations for sea cruising, sending native laborers aboard Saturn. Left chief steward Lynn and seaman Harbour onshore. Left alongside Saturn at 11.55 and stood to sea.

THOS. C. WOOD.

Noon to 4 P. M.

Weather pleasant: running to the westward from Port Ponce. At four o'clock had Cape Roco abeam, distant 5.2 miles. ———— ————, seaman, was put in double irons for safe keeping awaiting investigation by Captain.

GEORGE H. NORMAN, JR.

4 to 6 P. M.

Changed course to north at 5.30 and to N ½ E at six o'clock. Wind shifted from east south east to north by east. Weather threatening. Sighted Desacheo Island at 6.30 P. M.

A. M. PROCTER.

6 to 8 P. M.

Weather overcast and squally, changing latter part of watch; wind light from northeast. At 7.10 sighted Jiguero light two points on starboard bow. At 7.20 changed course to N N E ½ E., and saw loom of lights in the direction of Mayaguez.

THOS. C. WOOD.

8 P. M. to midnight.

Weather threatening and squally. At 8.42 had Jiguero light abeam distant 3.5 miles to eastward. At 10.30 changed course from N N E ½ E to east.

GEORGE H. NORMAN, JR.

STANDING ALONG COAST OF PUERTO RICO,

Midnight to 4 A. M.   *Wednesday*, August 17, 1898.

Stiff to fresh breezes from E ½ S. Standing along northern coast of Puerto Rico.

A. M. PROCTER.

4 to 8 A. M.

Daylight at 5.15. Made land on our starboard bow and beam. At 5.40 changed course to south and stood in to land. At 5.55 stood to eastward again and sighted wreck of steamer on east end of Cobras Island. At 6.40 made out forts at San Juan on our starboard bow. At eight o'clock the Morro of San Juan bore S by W on course E by N distant 3.5 miles, patent log reading 45.4; revolutions 78. Weather squally; swell moderate from eastward.

THOS. C. WOOD.

8 A. M. to noon.

Weather squally with hard rains. At 11.50 had Cape San Juan light abeam bearing S ¾ W, distant about five miles. Patrick Kay, ship's cook, 1st class, reported the expiration of his term of enlistment.

GEORGE H. NORMAN, JR.

Noon to 4 P. M.

Sighted Culebra Island at 12.30. Running on various courses from bearings. Sighted St. Thomas at 1.30. At end of watch, standing along coast of St. Thomas Island. Weather threatening.

A. M. PROCTER.

4 to 6 P. M.

Came to anchor in harbor of St. Thomas at five o'clock. Captain of the Port came on board. Following vessels were at anchor in the harbor: U. S. S. Cincinnati; a Danish man-of-war, the St. Thomas; and a Dutch man-of-war, the Friezland; also the Ville de Marseilles, tug Hercules and other shipping. Received official visits from Lieutenant Baron van Asbeck of the Friezland and Lieutenant Bastrup of the St. Thomas. Weather overcast and cloudy; fifteen fathoms of chain out.

J. T. EDSON.

6 to 8 P. M.

Weather pleasant; wind east; barometer steady.

THOS. C. WOOD.

8 P. M. to midnight.

Watch uneventful.   GEORGE H. NORMAN, JR.

8

St. Thomas, Danish West Indies, *Thursday*, August 18, 1898.
Midnight to 4 A. M.

At anchor in harbor of St. Thomas, D. W. I.; weather pleasant; watch uneventful.

A. M. Procter.

4 to 8 A. M.

An Italian steamer entered harbor at six o'clock, and went alongside of dock. Sent whaleboat ashore at 6.45. Scrubbing deck and washing paint work. Shore lighter came alongside at 7.10.

J. T. Edson.

8 A. M. to noon.

Lieutenant Huse called on Captains of Danish man-of-war St. Thomas and Dutch man-of-war Friezland and on Captain Chester of the Cincinnati. Executive Officer of Friezland called on Captain. Consul Van Horn came on board. Lieutenant Wood representing Captain called on Governor accompanied by consul Van Horn.

Thos. C. Wood.

Noon to 4 P. M.

An officer from the Friezland came on board at three o'clock bringing with him his commanding officer's card, returning call of our Executive Officer. Sent mail to Hercules about to sail north.

George H. Norman, Jr.

4 to 6 P. M.

Weather pleasant. Liberty men returning. Tug Hercules left port.

A. M. Procter.

6 to 8 P. M.

Weather pleasant; liberty men returning to ship. At eight o'clock eight men remained absent.

Thos. C. Wood.

8 P. M. to midnight.

Weather clear; wind in strong puffs from east north east.

Thos. C. Wood.

St. Thomas, Danish West Indies, *Friday*, August 19, 1898.
Midnight to 4 A. M.

Showers; watch uneventful.

GEORGE H. NORMAN, JR.

4 to 8 A. M.

Weather pleasant. Touched up rust spots outside of ship preparatory to painting ship. ——— ———, fireman 2nd class, returned on board eleven hours over time; ——— ———, machinist 1st class, returned on board thirteen hours overtime.

A. M. PROCTER.

8 A. M. to noon.

Water lighter came alongside at ten o'clock. Liberty party went ashore in shore boats at eleven to return at sundown. Cincinnati signalled to go alongside coaling lighter at 10.45. The Captain of the Danish cruiser came aboard at 11.05.

J. T. EDSON.

Noon to 4 P. M.

Got up anchor at one o'clock and lay alongside of coaling schooner Southard. Sent ashore for men to do the handling of coal. French steamer Ville de Marseilles left the harbor at two. Weather pleasant; wind moderating.

J. T. EDSON.

4 to 6 P. M.

Weather pleasant; lying alongside schooner Southard coaling ship and taking water.

GEORGE H. NORMAN, JR.

6 to 8 P. M.

Hauled away from coal schooner and anchored.

A. M. PROCTER.

8 P. M. to midnight.

Weather pleasant; watch uneventful.

J. T. EDSON.

St. Thomas, Danish West Indies, *Saturday*, August 20, 1898.
Midnight to 4 A. M.

Heavy and short rain squall about two o'clock; otherwise pleasant.

                     Thos. C. Wood.

4 to 8 A. M.

Weather pleasant. Called all hands at five. Preparing ship for painting. At 6.30, Dutch cruiser Friezland left the harbor.

                     George H. Norman, Jr.

8 A. M. to noon.

Anchored in harbor of Charlotte Amalia, St. Thomas, D. W. I., painting ship. Discharged ashes into ash lighter.

                     A. M. Procter.

Noon to 4 P. M.

Water lighter came alongside at 2.30 containing 5,500 gallons of water. ——— ———, machinist 2nd class, was placed in double irons for being under the influence of liquor while on duty. Light showers at three o'clock.

                     J. T. Edson.

4 to 6 P. M.

Weather pleasant, showery latter part of watch. Finished taking water from lighter. At 5.20 Captain Wainwright visited Captain Chester of Cincinnati. Paymaster brought on board from shore $2500.

                     Thos. C. Wood.

6 to 8 P. M.

Weather pleasant; watch uneventful.

                     George H. Norman, Jr.

8 P. M. to midnight.

Weather pleasant; watch uneventful.

                     A. M. Procter.

MAKING PASSAGE TO PONCE, PUERTO RICO,

Midnight to 4 A. M.   *Sunday*, August 21, 1898.

Weather clear and pleasant.   J. T. EDSON.

4 to 8 A. M.

Called all hands at 4.30. At five o'clock made preparations for
sea. At 5.20 signalled to Cincinnati for permission to get underway. At 5.40 called all hands up anchor and stood out of harbor
of St. Thomas. At 6.17 took departure from Flamingo Point, distant one-half mile bearing abeam. Set course S W ½ W for Ponce,
Puerto Rico; patent log reading 12.0. At 7.25 changed course to
S W to clear east point of Crab Island bearing one-half point on
starboard bow.   THOS. C. WOOD.

8 A. M. to noon.

Weather pleasant. At 8.07 with east point of Crab Island abeam
changed course from S W to W S W ½ W; log reading 31. At
9.10 had lighthouse on starboard beam (not charted), distant 2.3
miles. At 10.55 had Tupa light abeam, distant 3.2 miles, log reading 62.8.   GEORGE H. NORMAN, JR.

Noon to 4 P. M.

Underway for Ponce; passed San Domingo man-of-war. Sighted
Muertos Island at 1.30 P. M. Stood in to Ponce and anchored
on port quarter of Annapolis. Captain went aboard Annapolis and
returned. Paymaster Brown went ashore to settle bills. German
man-of-war got underway and stood to eastward.   A. M. PROCTER.

4 to 6 P. M.

Got up anchor and stood out of harbor for Guanica at 4.45. Patent log read 3.4 as we passed the red buoy at entrance. Came to
anchor at Guanica at 6.30. Weather clear and pleasant.   J. T. EDSON.

6 to 8 P. M.

At anchor in Guanica harbor. Signalled Puritan for permission
to remain till morning in order to obtain mail for Guantanamo.
Puritan, Terror and Amphitrite sent mail to us.   THOS. C. WOOD.

8 P. M. to midnight.

Weather pleasant until 11.30 when heavy rain and wind squall
struck harbor from southward and eastward.   GEORGE H. NORMAN, JR.

At Sea, *Monday*, August 22, 1898.

Midnight to 4 A. M.

Showers first part of watch.

A. M. Procter.

4 to 8 A. M.

Called all hands at five o'clock. Got up anchor at 5.40 and stood out; course set west at 6.30. Lighthouse on west end Puerto Rico abeam at 7.30, patent log 19.8 at 7.35. Course set N W ¼ W at 7.48. Transport No. 5 passed close to at 7.50. Weather squally; wind from east.

J. T. Edson.

8 A. M. to noon.

At eight o'clock Roco light, Puerto Rico, bore N 78° E., distant six miles. At 8.30 sighted Desacheo Island broad off starboard bow, distant about twenty-eight miles. At 9.45 Mono Island on port bow, distant about twelve miles. At 10.20 Desacheo Island bore abeam, distant ten miles. By order of Captain released from confinement ——— ———, machinist 2nd class.

Thos. C. Wood.

Noon to 4 P. M.

Sighted land on port bow at 1.45. At 2.30 changed course to N by W and ran seven miles to clear charted rocks. Ship was set to southward by current. At 3.10 changed course to N W, patent log reading 97.5. At end of watch Point Macao bore on the bow, patent log 6.

A. M. Procter.

4 to 6 P. M.

At 4.20 Point Macao bore abeam (S W), distant four miles. Weather overcast and cloudy with slight shower at 6.45.

J. T. Edson.

6 to 8 P. M.

Weather cloudy; wind light from northward and eastward. At 6.40 made Samana Point bearing N W ½ W, distant twenty miles. Sea smooth.

Thos. C. Wood.

8 P. M. to midnight.

Standing on course N W ¾ W; at 11.30 made out land on port beam. Changed course at 11.55 to N W ½ W.

A. M. Procter.

AT SEA, *Tuesday*, August 23, 1898.

Midnight to 4 A. M.

Course changed to W N W ½ W at 1.43, patent log reading 11.8. Weather squally.

J. T. EDSON.

4 to 8 A. M.

Weather cloudy; wind freshening from eastward. At daylight (5 A. M.) made Point Patella on port bow, distant about twenty miles.

THOS. C. WOOD.

8 A. M. to noon.

Changed course to west at eight o'clock. Island of Haiti in sight along port side, distant about twenty-five miles. Quarters for inspection at 9.30 followed by inspection of boat gear. Sighted Island of Tortuga two points on port bow at 1.55.

A. M. PROCTER.

Noon to 4 P. M.

Changed course to W by S at noon. Weather pleasant. The Island of Tortuga bears south, distant about five miles.

J. T. EDSON.

4 to 6 P. M.

Weather pleasant. At 4.51 west point of Tortuga Island bore abeam, distant four miles.

THOS. C. WOOD.

6 to 8 P. M.

Weather pleasant. Coast of St. Nicholas Mole bore S S W at 1.10, distant about twenty miles.

A. M. PROCTER.

8 P. M. to midnight.

Weather clear and pleasant. Course W by S, making 78 revolutions or a little under eight knots. No lights or land sighted.

J. T. EDSON.

AT SEA, *Wednesday*, August 24, 1898.

Midnight to 4 A. M.

Weather pleasant; wind light, sea smooth. At 12.30 sighted what was thought to be Maysi light bearing north north west about seventeen miles distant. Sighted several steamer lights during watch, the vessels apparently standing to the northward and eastward.

THOS. C. WOOD.

4 to 8 A. M.

At daylight made out Cuban coast on starboard beam. Changed course to W N W and ran in towards coast at 5.10 A. M. At 6.45 headed W ½ N. Standing towards Guantanamo at slow speed at end of watch.

A. M. PROCTER.

8 A. M. to noon.

Dropped anchor in Guantanamo at 8.45 A. M. Captain Brownson came on board. Captain Wainwright visited flagship Newark. Anchored in the harbor are the Newark, Marblehead, New Orleans, Dixie, Yankee, Glacier, Vulcan, Vixen, Hist, Kanawah, Fern, and Solace. The Niagara with Saturn in tow entered harbor at 10.45. George Lynn, wardroom steward, and Wm. A. Harbour, master-at-arms 3rd class, reported on board at 11.45 from the Niagara.

J. T. EDSON.

Noon to 4 P. M.

At anchor at Guantanamo. At two o'clock, Fern got underway and stood out to sea. Sent mail to Dixie at 3.15. At four o'clock, Dixie got underway and stood out of harbor for New York, followed at 4.10 by Solace. At 1.20 sent Lawerence, ordinary seaman, sick, to Solace.

THOS. C. WOOD.

4 to 6 P. M.

Guard boat came alongside bringing official mail. William Thomas, seaman, was transferred to this ship from Marblehead with papers and effects.

A. M. PROCTER.

6 to 8 P. M.

At anchor at Guantanamo; weather pleasant. Received signal from Flagship to get coal to-morrow from Leonidas.

J. T. EDSON.

8 P. M. to midnight.

Weather pleasant and calm; visiting party returned from New Orleans at 8.15.

THOS. C. WOOD.

GUANTANAMO, CUBA, *Thursday*, August 25, 1898.

Midnight to 4 A. M.

Weather pleasant; watch uneventful.

A. M. PROCTER.

4 to 8 A. M.

Crew engaged in scrubbing clothes. Rigged in boats and port boom and made preparation for going alongside Sterling for water. Weather pleasant. At 7.05 in response to signal, sent Steward to Glacier after stores. Returned at 8.05.

J. T. EDSON.

8 A. M. to noon.

Weather pleasant. At 8.20 moved up alongside the Sterling to get water; finding however, it was too salty, decided not to take it aboard and moved away and anchored. Usual routine of ship observed. Got aboard provisions, meat, etc.

THOS. C. WOOD.

Noon to 4 P. M.

Anchored in Guantanamo bay; weather pleasant. Struck below all ammunition and unloaded pistol belts. Commander-in-Chief flying his pennant visited Sterling, Saturn, and Niagara.

A. M. PROCTER.

4 to 6 P. M.

At five o'clock received visit from Commodore Watson and Captain Goodrich. Weather pleasant. Board of Survey condemned and threw overboard one-half barrel of musty flour.

THOS. C. WOOD.

6 to 8 P. M.

No change in weather; parties from New Orleans, Glacier, Hist, and Sterling visited the ship and returned at eight o'clock.

THOS. C. WOOD.

8 P. M. to midnight.

Weather calm and pleasant; at 10.20 Montgomery got underway and left the harbor with Commodore Watson.

THOS. C. WOOD.

GUANTANAMO, CUBA, *Friday,* August 26, 1898.
Midnight to 4 A. M.

No change in weather conditions; watch quiet.

THOS. C. WOOD.

4 to 8 A. M.

At six o'clock sent cutter to Glacier for provisions and ice with
Chief Pay Yeoman and Steward. Made preparations for coaling
ship, and went alongside Leonidas for coal at seven.

THOS. C. WOOD.

8 A. M. to noon.

Began coaling ship from Leonidas at 8.15. Coal hot and containing some sulphur.

A. M. PROCTER.

Noon to 4 P. M.

Continued to coal ship.

A. M. PROCTER.

4 to 8 P. M.

Finished coaling ship having taken on board seventy-one tons of
bituminous coal. Hauled out from collier and anchored; cleaned
ship; and sent men in swimming.

A. M. PROCTER.

8 P. M. to midnight.

Captain McCalla called on board; Captain Young and officers of
Hist called on board. Allowed fires to die out in boiler A.

A. M. PROCTER.

GUANTANAMO, CUBA, *Saturday*, August 27, 1898.

Midnight to 4 A. M.

Watch uneventful; weather pleasant.

A. M. PROCTER.

4 to 8 A. M.

Sent boat to Glacier for supplies; began general cleaning day.

A. M. PROCTER.

8 A. M. to noon.

Copy of specifications in the case of ——— ———, seaman, to be tried by Summary Court Martial were delivered to the accused. Cutter with Paymaster and Steward sent to Glacier for stores at nine o'clock; returned at 10.45. Unloaded cutter and sent it to Niagara with several packages of coal bags. Crew engaged until 11.15 in scrubbing deck and cleaning ship.

J. T. EDSON.

Noon to 4 P. M.

Sterling, Saturn, and Niagara got up anchor and stood out to sea at noon. At one o'clock the Sterling signalled she was aground. Nashville was ordered by Flagship to get underway and assist Sterling. Nashville returned to her anchorage at three. Captain Wainwright and Lieutenant Huse left the ship at 3.30.

J. T. EDSON.

4 to 6 P. M.

First whaler returned at 5.20. A Norwegian merchant steamer came out of inner harbor at 5.30 and stood out to sea. At five o'clock the Captain and Lieutenant Huse returned. Served out clean hammocks at evening quarters.

J. T. EDSON.

6 to 8 P. M.

The Montgomery entered the harbor at 7.30 P. M. Weather pleasant.

J. T. EDSON.

8 P. M. to midnight.

Montgomery came to anchor at eight o'clock. Weather clear.

J. T. EDSON.

GUANTANAMO, CUBA, *Sunday*, August 28, 1898.

Midnight to 4 A. M.

No change in weather; watch uneventful.

                                                  J. T. EDSON.

4 to 8 A. M.

Crew engaged in scrubbing deck and getting ready for inspection. At 6.35 sent cutter to Glacier for provisions. At 6.45 steamship Milly Brooks, showing no colors, entered the harbor from the eastward and stood in to the inner harbor. She was signalled, " Heave to, I have important news for you " by the Flagship, but made no response.

                                                  J. T. EDSON.

8 A. M. to noon.

Usual Sunday inspection at ten o'clock by Lieutenant-Commander Wainwright. Commanding. Petty officers were granted permission to go sailing in first whaleboat and remain until 5 P. M.

                                                  THOS. C. WOOD.

Noon to 4 P. M.

Weather pleasant. Visiting party to Newark returned at four o'clock. Officers of other vessels visited this ship.

                                                  THOS. C. WOOD.

4 to 6 P. M.

Weather unchanged; watch uneventful.

                                                  THOS. C. WOOD.

6 to 8 P. M.

Weather pleasant. Lieutenant Huse and Chief Engineer McElroy visited Newark. At 7.50, Hist got underway for Kingston, Jamaica, and passed under our stern bound out.

                                                  THOS. C. WOOD.

8 P. M. to midnight.

Steamer Milly Brooks from Caymanera passed down river and out of harbor. At ten o'clock Montgomery got underway and stood out to sea bound for Guanica, Puerto Rico. Weather pleasant.

                                                  THOS. C. WOOD.

GUANTANAMO, CUBA, *Monday*, August 29, 1898.

Midnight to 4 A. M.

Weather pleasant; watch uneventful.

THOS. C. WOOD.

4 to 8 A. M.

Called all hands at five o'clock. Sent cutter with Paymaster Brown and Steward to Glacier for ice and meat. Scrubbed hammocks. At six o'clock, Nashville got up anchor and went alongside Leonidas for coal.

THOS. C. WOOD.

8 A. M. to noon.

Went alongside Vulcan at 9.55. Summary Court Martial on ———— ————, seaman, convened at 10.05. Testimony being all in, court adjourned at 11.20 and prisoner was confined in double irons.

GEORGE H. NORMAN, JR.

Noon to 4 P. M.

Went alongside of Leonidas at noon and took onboard twenty-one tons of coal, casting off at 4.30. Received from U. S. S. Scorpion George Picton, seaman, for passage home.

GEORGE H. NORMAN, JR.

4 to 6 P. M.

Received from U. S. S. Vulcan F. F. Shipp, Chief Yeoman, for passage home.

GEORGE H. NORMAN, JR.

6 to 8 P. M.

Naval Cadets H. M. Gleason, C. A. Weichart, J. H. Tomb, and H. G. Sparrow reported on board from U. S. S. Marblehead. Weighed anchor at 6.05 and stood out of bay. With an offing of about two miles made course east, patent log reading 2.6.

GEORGE H. NORMAN, JR.

8 P. M. to midnight.

Weather pleasant; bright moonlight. Sighted small sloop inshore at nine o'clock.

A. M. PROCTER.

Midnight to 4 A. M.   AT SEA, *Tuesday*, August 30, 1898.

Changed course to N E by N at 1.04, patent log reading 56.6. Maysi light sighted at 2.04 bearing three points on port bow. At 3.08 changed course to N by W, patent log reading 72. Maysi light was abeam at 3.50, distant four miles, patent log 78.8. Weather pleasant, wind northeast.

J. T. EDSON.

4 to 8 A. M.

Clear and pleasant; gentle breezes from northeast by east; barometer rising. At 5.30 put second boiler into use and increased number of revolutions to 100 per minute. At sunrise, sun bore N 83° E, course N by W. At 6.15 passed steamer bound south.

THOS. C. WOOD.

8 A. M. to noon.

Weather pleasant. Course at eight o'clock, N by W; changed course at nine to N ½ W. At 11.10 sighted steamer on starboard beam standing to southward. At 11.30 sighted steamer on starboard bow and at 11.50 another steamer on starboard bow both standing to the southward.

GEORGE H. NORMAN, JR.

Noon to 4 P. M.

Weather pleasant; gentle breezes from east; barometer falling slowly. At two o'clock sighted Mira Por Vos Islets. At 2.15 changed course to N N E, patent log 86.7. At 2.20 changed course to N by W, patent log 87.6. At 2.54 changed course to N ¾ E, patent log 95.

A. M. PROCTER.

4 to 6 P. M.

Made Fortune Island at 4.45, bearing N 45° E, patent log reading 15.7; abeam at 5.45, patent log reading 26. Changed course to N by E ½ E at 6.30, patent log 34.6. Bow bearing of Bird Rock light taken at 6.30, patent log 34.6.

J. T. EDSON.

6 to 8 P. M.

Weather pleasant; barometer steady, rising slightly. At 7.17 Bird Island light bore abeam nine miles distant.

THOS. C. WOOD.

8 P. M. to midnight.

Weather pleasant; at 11.50 sighted a white light in the horizon half a point on starboard bow. Course N by E ½ E.

GEORGE H. NORMAN, JR.

AT SEA, *Wednesday*, August 31, 1898.

Midnight to 4 A. M.

Weather pleasant; bright moonlight. Passed three steamers, all heading about S by W. At 1.40 sighted Watling Island light. At 2.10 Watling Island light abeam, distant ten miles, patent log 16.7; changed course to N ½ E.

A. M. PROCTER.

4 to 8 A. M.

Weather pleasant; light air from northeast. Crew engaged in scrubbing decks. Course N ½ E.

J. T. EDSON.

8 A. M. to noon.

Weather warm and pleasant; sea smooth. At 10.30 changed course to north. Latitude at noon N 25° 43'.

THOS. C. WOOD.

Noon to 4 P. M.

Weather pleasant with occasional showers. Course north. Watch uneventful.

GEORGE H. NORMAN, JR.

4 to 6 P. M.

Weather pleasant. Called all hands to muster at five o'clock and published the finding and sentence of a Summary Court Martial in the case of —— ——, seaman as follows: Said —— —— to be disrated to ordinary seaman and to loose fifty dollars' pay. —— ——, was released from arrest and restored to duty.

A. M. PROCTER.

6 to 8 P. M.

Weather clear and pleasant; light breeze from southeast. Watch uneventful.

J. T. EDSON.

8 P. M. to midnight.

Weather pleasant; occasional light showers: bright moonlight. Swell moderate from southward and eastward; barometer steady.

THOS. C. WOOD.

AT SEA, *Thursday*, September 1, 1898.

Midnight to 4 A. M.

Weather pleasant; bright moonlight; course north. Watch uneventful.

GEORGE H. NORMAN, JR.

4 to 8 A. M.

Weather pleasant; barometer rising. Executed morning orders.

A. M. PROCTER.

8 A. M. to noon.

Clear and pleasant; gentle breezes from east south east and east by north. Barometer rising. Rain squalls to windward. Watch uneventful. Aired bedding during watch.

J. T. EDSON.

Noon to 4 P. M.

Clear and pleasant; barometer steady; light breezes from eastward. Piped down aired bedding and scrubbed and washed clothes at seven bells. The Officer of the Deck condemned about five pounds of bread and five pounds of flour, belonging to No. 1 mess.

THOS. C. WOOD.

4 to 6 P. M.

Weather pleasant; at five o'clock changed course from north to north one-quarter east.

GEORGE H. NORMAN, JR.

6 to 8 P. M.

Weather pleasant; barometer high.

A. M. PROCTER.

8 P. M. to midnight.

Weather pleasant; bright full moon. Engines making 92 revolutions. At 11.50 a green light was reported by forecastle lookout one point on starboard bow. This light was also seen by Quartermaster but soon disappeared.

J. T. EDSON.

At Sea, *Friday*, September 2, 1898.

Midnight to 4 A. M.

Weather clear and pleasant; bright moonlight; sea smooth. No lights or vessels sighted during watch. Revolutions 92.3; distance run during watch, 39.4 knots.

THOS. C. WOOD.

4 to 8 A. M.

Weather bright; light airs from eastward and flat calms; sea glassy; watch uneventful.

GEORGE H. NORMAN, JR.

8 A. M. to noon.

Weather calm and warm; barometer high. Overhauled battery. Changed course at noon to N by E, patent log 76.3.

A. M. PROCTER.

Noon to 4 P. M.

Weather clear and pleasant; light air from westward. Crew engaged on boat gear. A steamer sighted on port bow headed westward.

J. T. EDSON.

4 to 6 P. M.

Weather clear and pleasant. Vessel reported in previous watch proved to be the U. S. S. Annapolis with whom we exchanged signals, asking, " Where are you bound? "

THOS. C. WOOD.

6 to 8 P. M.

Weather pleasant; watch uneventful.

GEORGE H. NORMAN, JR.

8 P. M. to midnight

Weather pleasant; bright moonlight; freshening breeze. Sighted Diamond Shoal lightship at 9.55. Changed course at ten o'clock to N N E ½ E, Diamond Shoal light abeam.

A. M. PROCTER.

9

AT SEA, *Saturday*, September 3, 1898.

Midnight to 4 A. M.

Weather clear and pleasant; bright moonlight; wind from the west and gradually freshening. No sail or lights sighted. A sounding taken at four o'clock showed no bottom at sixty fathoms.

J. T. EDSON.

4 to 8 A. M.

Weather clear; wind freshening from westward; barometer slowly falling. No change of course during watch and no sails or lights sighted.

THOS. C. WOOD.

8 A. M. to noon.

Weather pleasant; course changed at nine o'clock from N N E ¼ E to N N E.

GEORGE H. NORMAN, JR.

Noon to 4 P. M.

Changed course at one o'clock to N by E ½ E. Passed two steamers heading to southward and one brigantine heading to northward. Finished scrubbing.

A. M. PROCTER.

4 to 6 P. M.

At 4.45 changed course to N by E, patent log 62.7. A sounding taken at five o'clock showed twenty-five fathoms. Barometer still falling about one hundredth each hour.

J. T. EDSON.

6 to 8 P. M.

Weather pleasant; wind moderating. At 7.40 sighted Five Fathom lightship three points on the port bow. Moon rose at 7.50.

THOS. C. WOOD.

8 P. M. to midnight.

Weather pleasant. At nine passed a tow standing to southward. At 9.27 changed course from N by E to north; at 9.40 to N E ½ N; at 9.50 to N E and passed a tow bound north. At 11.35 had Absecom light on port bow, patent log reading 31.1. At midnight, sounding gave fifteen fathoms.

GEORGE H. NORMAN, JR.

AT SEA, *Sunday*, September 4, 1898.

Midnight to 4 A. M.

Weather cool and pleasant; barometer steady. Absecom light abeam at 12.57, distant twelve miles. Sighted Tucker's Island light at 1.15, and Barnegat light at 1.30. Barnegat light was abeam at 3.51, distant thirteen miles. Passed five steamers and one barge all heading to southward. Took sounding at four o'clock in fifteen fathoms.

A. M. PROCTER.

4 to 8 A. M.

Course changed to north at 5.40, patent log 90.9. A sounding taken at six o'clock showed sixteen fathoms. Land sighted on port bow at 6.15 A. M. Made the Highlands at 6.45. Weather pleasant; numerous sails in sight. The Jersey shore in plain sight along the port side.

J. T. EDSON.

8 A. M. to noon.

Clear and pleasant. Made Scotland lightship and ran in through Swash Channel. Made ship's name K J S V by international signal code to Sandy Hook station. Station signalled F D G S (welcome); replied D W Q P (thanks). Many steamers saluted with whistles or cheered as we passed up the bay. Stopped a few minutes at Quarantine for pratique and went on to Tompkinsville. Exchanged numbers with Indiana and got permission to anchor. Anchored in six fathoms of water with thirty fathoms of chain. Ships of North Atlantic squadron at anchorage cheered as we steamed past.

HARRY P. HUSE.

AFTER ONE HUNDRED DAYS.

The Gloucester was back in New York harbor after an absence of just one hundred days. During that time she had been in action at Aguadores, Santiago, Guanica, and Arroyo, and had facilitated the landing of troops in Puerto Rico by cutting out lighters at Guanica and Port Ponce. With a few exceptions her complement was unchanged: J. W. Williams, coal-passer, was sent home suffering from injuries incurred before enlistment; W. H. Lawerence, ordinary seaman, was transferred to the Solace with an obstinate case of fever which, however, soon left him on reaching a better climate; A. Jaggi, coxswain, was transferred to the Newark at his own request just before coming north; Mr. Procter, who had been left in the navy-yard hospital, rejoined the ship in time to be in the battle of July 3rd; and William Thomas, seaman, was transferred from the Marblehead August 24th.

The rest of the cruise was uneventful. On the 10th of September, the Gloucester sailed for Boston where she remained till the 15th, going thence to visit the picturesque old New England city after which she was called. There had been much disgust felt by the people of Gloucester when that name was given to the yacht Corsair, and perhaps they felt some jealousy that Castine, Machias, and above all, Marblehead, should be the sponsors of larger vessels. But now they extended the most enthusiastic welcome to the little yacht. The whole population seemed to have turned out to do her honor, and amid the screaming of whistles, the waving of flags, and the cheers of the thousands afloat or lining the water-front, the Gloucester came to anchor in the inner harbor.

Everything was done by the town's people to make the visit a pleasant and a memorable one. A stay of three days, during which the ship was thrown open to the public, enabled all to come on board who cared to do so; and thousands availed themselves of the opportunity. As a mark of appreciation of the hospitality of the city the national ensign taken ashore by the landing party at Guanica, and therefore the first United States flag hoisted on the island of Puerto Rico, was presented to the Mayor.

The Gloucester returned to Boston September 18th and remained there till the 25th. On the way to New York, thick weather induced the captain to anchor overnight in Provincetown. On the 27th she was back at the anchorage off Tompkinsville; and on the 1st of October she was underway for Annapolis, where the superintendent had requested she be sent for the use of the cadets.

The citizens of Philadelphia asked that the Gloucester be present at their great peace festival, and the request was granted. In the procession, the "Gloucesters" were accorded a prominent place, and their appearance always called forth cheers and enthusiastic applause.

The ship then returned to Annapolis and moored alongside the Santee's wharf. By this time nearly all the original officers had been detached, and in a few days Ensign Thomae, who had joined the ship in Boston as the relief of Lieutenant Wood, was left in command with just sufficient men to keep the ship in order. The pennant has not been hauled down, and the colors that flew at Aguadores, Santiago, Guanica, Ponce, and Arroyo, are still hoisted every morning at eight o'clock. Lieutenant-Commander Wainwright from the deck of the old Santee can look across the wharf at his late command and still exercises authority over her as the Officer in Charge of Ships, U. S. Naval Academy, and Assistant Engineer Procter, who has relieved Ensign Thomae, reports daily to his former captain the condition of his ship.

U. S. Naval Academy, Annapolis, Md.
    December, 1898.

Roster of expeditions of which no special report was made, but which are recorded in the log.

Landing party at Banes, Cuba, June 6th, 1898:
Lieutenant A. H. Dutton, commanding.
Whitelock, Quentin, Tierney, Loehrs, Macklin, Daly.

Landing party to protect Spanish prisoners, July 3rd, 1898:
Lieutenant George H. Norman, commanding.
Bond, Thompson, Rozzle, Lykke, Lewis, Mulcahey, Hillman, Halverson, Noble, Dahl, Collin, Loehrs.

Cutting-out expedition at Port Ponce, July 27th, 1898:
Cutter: Lieutenant Harry P. Huse, commanding.
Assistant Paymaster Alexander Brown.
Keller (Cox), Chipman, Englert, Davis, Wirtane, Murphy, Rozzle, Halverson, Tierney, Hillman, Loehrs, Kleinkopf, Brown.
Whaler: Lieutenant George H. Norman, Jr., commanding.
Bond, Dahl (Cox), Harbour, Kastell, Thompson, Whetton, Lykke, Mulcahey.

REPORTS OF
CAPTAIN WAINWRIGHT AND OFFICERS, ON THE
BATTLE OF JULY 3, 1898

STATIONS IN BATTLE, JULY 3, 1898.

On the Bridge:

Lieutenant-Commander Wainwright, Captain.
Lieutenant Huse, Executive and Navigator.
Assistant Engineer Procter, Aid.
Chief Quartermaster Bechtold, Signalman.
Quartermaster Green, at the Wheel.
Quartermaster Noble, Orderly.

First Division, Starboard:

Lieutenant Norman, Commanding Division.
Gun No. 1, 3-pounder rapid-fire, Bond.
Gun No. 3, 6-pounder rapid-fire, Whitelock.
Gun No. 5, 6-pounder rapid-fire, Dahl.

First Division, Port:

Ensign Edson, Commanding Division.
Assistant Surgeon Bransford, Volunteer (fired No. 4 gun).
Gun No. 2, 6-pounder rapid-fire, Keller.
Gun No. 4, 6-pounder rapid-fire, Lynch.
In charge of ammunition supply, forward, Meehan.

Second Division:

Lieutenant Wood, Commanding Division.
Gun No. 8, 3-pounder rapid-fire, fired by Lieut. Wood.
Gun No. 9, 3-pounder rapid-fire, Jaggi.
Gun No. 10, 3-pounder rapid-fire, Lacy.
In charge of ammunition supply, aft, Bee.

Auxiliary Division:

Assistant Paymaster Brown, Commanding Division.
Gun No. 6, 6-millimetre Colt automatic, Mr. Brown.
Gun No. 7, 6-millimetre Colt automatic, Chipman.

Engineer Division:

Passed Assistant Engineer McElroy, Commanding Division.
At reversing gear and throttle, Jennings.
At annunciator, Johanson.
Overseeing oilers, etc., Hare.
In charge of fire-room, McKeon.

In charge of dynamo, Graves.

REPORT OF LIEUTENANT-COMMANDER
RICHARD WAINWRIGHT, CAPTAIN.

U. S. S. Gloucester,
Off Santiago de Cuba, Cuba.
July 6, 1898.

SIR:—

1. I have the honor to report that at the battle of Santiago on July 3rd, the officers and crew of the Gloucester were uninjured and the vessel was not injured in hull or machinery, the battery only requiring some slight overhauling. It is now in excellent condition.

2. I enclose herewith a copy of the report of the executive officer, made in compliance with paragraph 525, page 110, Naval Regulations, which report I believe to be correct in all particulars. I also enclose copies of the reports of the several officers which may prove valuable for future reference.

3. It was the plain duty of the Gloucester to look after the destroyers, and she was held back, gaining steam, until they appeared at the entrance. The Indiana poured in a hot fire from all her secondary battery upon the destroyers, but Captain Taylor's signal, " Gunboats close in," gave security that we would not be fired on by our own ships. Until the leading destroyer was injured, our course was converging necessarily; but as soon as she slackened her speed, we headed directly for both vessels, firing both port and starboard batteries as the occasion offered.

4. All the officers and nearly all the men deserve my highest praise during the action. The escape of the Gloucester was due mainly to the accuracy and rapidity of the

fire. The efficiency of this fire, as well as that of the ship generally, was largely due to the intelligent and unremitting efforts of the executive officer, Lieutenant Harry P. Huse. The result is more to his credit when it is remembered that a large proportion of the officers and men were untrained when the Gloucester was commissioned. Throughout the action he was on the bridge and carried out my orders with great coolness. That we were able to close in with the destroyers, and until we did so they were not seriously injured, was largely due to the skill and constant attention of Passed Assistant Engineer George H. McElroy. The blowers were put on and the speed increased to seventeen knots without causing a tube to leak or a brass to heat. Lieutenant Thomas C. Wood, Lieutenant George H. Norman, Jr., and Ensign John T. Edson, not only controlled the fire of the guns in their divisions and prevented waste of ammunition, but they also did some excellent shooting themselves. Acting Assistant Surgeon J. F. Bransford took charge of one of the guns and fired it himself occasionally. Acting Assistant Paymaster Alexander H. Brown had charge of the two Colt guns, firing one himself, and they did excellent work. Assistant Engineer A. M. Procter carried my orders from the bridge and occasionally fired a gun when I found it was not being served quite satisfactorily. All were cool and active at a time when they could have had but little hope of escaping uninjured.

5. Lieutenants Wood and Norman, Ensign Edson, and Assistant Engineer Procter were in charge of the boats engaged in saving life. They all risked their lives repeatedly in boarding and remaining near the two destroyers and the two armored cruisers when their guns were being discharged by the heat and their magazines and boilers were exploding. They also showed great skill in landing and taking off the prisoners through the surf.

6. Of the men mentioned in the several reports I would call especial attention to John Bond, Chief Boatswain's Mate. He would have been recommended to the Department for promotion prior to his gallant conduct during the action of July 3rd. I would also recommend to your attention Robert P. Jennings, Chief Machinist, mentioned in the report of Mr. McElroy. I believe it would have a good effect to recognize the skill of the men and the danger incurred by the Engineer's force. I would also recommend that the acting appointments of those men mentioned by the officers in their reports be made permanent.

7. The wounded and exhausted prisoners were well and skillfully tended by Assistant Surgeon Bransford, assisted by Ensign Edson, who is also a surgeon.

8. The Admiral, his officers, and men, were treated with all consideration and care possible. They were fed and clothed as far as our limited means would permit.

Very respectfully,

RICHARD WAINWRIGHT,
Lieutenant-Commander, U. S. N., Commanding.

To the Commander-in-Chief, U. S. Naval Force,
    North Atlantic Station.

REPORT OF LIEUTENANT HARRY P. HUSE, EXECUTIVE OFFICER.

U. S. S. Gloucester,
Off Santiago de Cuba,
July 4, 1898.

SIR:—

1. I have the honor to submit the following report on the battle of Santiago, July 3rd, 1898.

2. At 9.43 A. M. the Gloucester, then being about 3000 yards southeast of Morro, the officer of the deck reported that the Spanish fleet was coming out of Santiago. All hands were called to general quarters. You came on the bridge and I took the deck. Fire was opened at 3500 yards from the after guns (3-pounder rapid-fire); and, as they were brought to bear, from the bow gun (3-pounder rapid-fire) and the starboard guns forward (6-pounder rapid-fire). The fire-room blowers were started, and, turning to starboard, the range was decreased to 3000 yards. Four Spanish cruisers came out in column and stood to the westward close in shore. In the belief that the two torpedo destroyers known to be in the harbor would come out, you directed me to slow down and wait for them, keeping up a deliberate fire on the cruisers from the port battery. There was no other gunboat with the fleet at the time, and the battleships Iowa, Indiana, Oregon and Texas and the armored cruiser Brooklyn were engaged with the four Spanish vessels, Cristobal Colon, Oquendo, Viscaya and Infanta Maria Teresa (flag), all standing to the westward under full head of steam. The forts on shore kept up a slow fire throughout the action till it was evident to them that our boats were being used to rescue Spanish seamen, when their fire ceased.

3. When the larger vessels were well clear and the rear one about 1500 yards to the westward of Morro, the destroyers Pluton and Furor came out and followed in their wake. At once we opened rapid fire on them from the starboard battery at a range of 2500 yards, and the engines were run at full speed, the ship heading about north north west. Presently signal was made from the Indiana, "Gunboats will advance." After this signal it appeared that the fight between this ship and the two apparently uninjured destroyers was a thing apart from the battle in which the larger ships were engaged. The starboard forward guns (one 3-pounder and two 6-pounder rapid-fire) were turned on the leading vessel, the Pluton, while the starboard after gun and the stern gun (both 3-pounder rapid-fire) were aimed at the Furor. The speed of the Gloucester was gradually increased to over 17 knots, and then we were slowly overhauling the torpedo-destroyers and closing in towards them. The fire from both sides was vigorous, but while many shots struck the water close alongside or went whistling over our heads, we were not hit once during the whole engagement. This is the more remarkable as the monotonous reports of an automatic gun could be heard after the 2500 yard range was passed and the zone of fire could be distinctly traced by a line of splashes describing accurately the length of the ship and gradually approaching it. But at a distance variously estimated from ten to fifty yards, the automatic fire suddenly ceased. It was afterwards found to be from a 1-pounder Maxim, and the execution aboard would have been terrible during the few minutes that must have elapsed before the ship was sunk had the fire reached us. Meanwhile, the service of our own guns was excellent, and at a range of twelve hundred yards, the two 6-millimetre automatic Colt rifles opened on the enemy. The Pluton had now (about 10.15) slackened her speed, showing

evident signs of distress, and our fire was concentrated on the Furor. The range was decreased to 600 yards, and at this distance the majority of shots appeared to strike. The Pluton was run on the rocks about four miles west of Morro and blew up. Our crew cheered at the sight of the explosion. The Furor soon commenced to describe circles with a starboard helm, her firing ceased, and it became apparent that she was disabled. A white rag was waved from forward and we stopped firing. Lieutenants Wood and Norman and Assistant Engineer Procter were sent to rescue the crews and to see if the prizes could be saved. They found a horrible state of affairs on the Furor. The vessel was a perfect shambles. As she was on fire and burning rapidly, they took off the living and then rescued all they could find in the water and on the beach. The Pluton was among the rocks in the surf and could not be boarded, but her crew had made their way ashore or were adrift on life-buoys and wreckage. These were all taken on board. I have since learned that the New York passed a number of men in the water who had doubtless jumped overboard from the destroyers to escape our fire. All these were probably drowned.

4. While this work was going on, several explosions took place on the Furor; and presently, about 11.30, she threw her bow in the air and turning to port slowly sunk in deep water.

5. The following were rescued from the destroyers and are believed to be the only survivors:

> *Furor—*
> Commander Carlier,
> Lieutenant Arderius (badly wounded),
> 3 petty officers,
> 14 enlisted men.
> 19 total.

Pluton—
   Commander Vasquez,
   Lieutenant Boado,
   4 petty officers,
   20 enlisted men,
   26 total.

6. It was stated by Commander Carlier that the total complement of the Furor was 64, officers and enlisted men. That of the Pluton was doubtless the same.

7. While one of our boats was still ashore, seeing heavy clouds of smoke behind the next point, the ship was moved in that direction, the men being at quarters and everything in readiness for further action. On rounding the point, two men-of-war were found on the beach burning fiercely aft, the majority of the crew being crowded on the forecastles and unable apparently to reach land, only 200 yards away. Our boats under Lieutenant Norman and Ensign Edson put off to the nearer vessel, which proved to be the flagship Infanta Maria Teresa, and rescued all on board by landing them through the surf. Lieutenant Norman formally received the surrender of the Commander-in-Chief and all his officers and men present; and, as soon as all hands had been transferred ashore, brought on board this ship all the higher officers including the Admiral. Lieutenant Wood meanwhile rescued the remaining survivors on board the Oquendo, the second of the burning vessels.

8. The Spanish officers not feeling that the prisoners on shore were secured from attack by Cuban partisans, by your orders I directed Lieutenant Norman to land with a small force, establish a camp on shore, and hoist the United States flag over it. He took with him all the rations that could be spared from the stores on board.

9. There were several incidents of interest that have not been related in this report which I will refer to briefly. The colors of the Furor and Oquendo were brought on

10

board by Lieutenant Wood and the colors of the Pluton by Mr. Procter.

10. The Flagship New York while hastening from Siboney to join in the general action saw the Gloucester close to her two disabled antagonists and cheered her as she went by.

11. The Indiana made the general signal to the Gloucester, "Congratulations."

12. During the night, the ship being on blockading station, the Assistant Chief of Staff hailed us from a torpedoboat and after inquiring about our casualties, added: "The Admiral admired your splendid work."

13. By order of Captain Evans, the Admiral and his staff were transferred from this ship to the Iowa, all other unwounded prisoners were sent to the Indiana, and the twenty-two wounded were taken to Siboney and put on board the army hospital steamer Olivette. One wounded prisoner died on board and was buried at sea on the way back from Siboney.

14. A comparison of the armaments of the contending vessels is interesting:

>*Furor—*
>Length, 320 feet; displacement, 370 tons.
>Armament: 2 14-pds. R. F. G.
>2 6-pdr. R. F. G.
>2 1-pdr. Maxim automatic.
>2 14-inch torpedo tubes.
>Complement, 67.

>*Pluton—*
>The same.

>*Gloucester* (late the yacht *Corsair*, N. Y. Y. C.)
>Length, 241 feet; displacement, 800 tons.
>Armament: 4 6-pdr. R. F. G.
>4 3-pdr. R. F. G.
>2 6-mm. Colt automatic.
>Complement, 93.

15. The action was a remarkable one. The *matériel* of the enemy was superior in every respect; and yet, having destroyed two vessels either one of which would have been a fair match for this ship, and inflicted terrible loss to their *personnel*, I have to report not one casualty. This result I attribute wholly to the accuracy and rapidity of our fire, which made the proper service of the guns on board the Spanish ships utterly impossible. In this opinion I am borne out by the statements of our prisoners, who commented on the awful destructiveness of our fire and spoke of their unsuccessful efforts to use their torpedoes, the crews being swept away repeatedly by bursting shell. They also referred to the deadly effect of the Colt automatic gun (6 mm.), and said that the projectiles from these passed clean through the vessels.

16. While I may not say that any of our officers surpassed the others in gallantry or efficiency, I cannot refrain from enumerating them and again calling your attention to their good services.

17. Lieutenant Wood in command of the after division performed his duties in action with great energy and efficiency. After the action was over he bent all his efforts to saving life, and it is due to his efforts and those of Lieutenant Norman, Ensign Edson, and Assistant Engineer Procter, ably seconded by the men under their commands, that over 600 officers and men were rescued from drowning.

18. Lieutenant Norman showed qualities during and after the action that indicate unusual fitness for a naval career. It fell to this young officer to receive the surrender of the Spanish Commander-in-Chief after having rescued him from his burning flagship.

19. Ensign Edson, although a greater part of his life has been spent in sedentary studies and pursuits, showed the result of his early training at the Naval Academy and in the

Navy by the manner in which he fought his guns during the action. The skill and ability he showed in handling his boat 'in the surf in the work of rescue excited the admiration of the seamen under his command. On his return to the ship, Mr. Edson, who is a man of mature age and a surgeon in high standing in New York, turned his attention to the wounded prisoners and seconded the efforts of Assistant Surgeon Bransford.

20. Passed Assistant Engineer McElroy did not come under my notice during the action; but the great speed developed by this ship when overtaking the enemy, the promptness with which this speed was attained from a condition of inaction, the intelligent response to signals from the bridge, all indicate the excellent condition of the discipline in his division and the material under his immediate charge. Assistant Paymaster Brown did effective service with his division of two 6 mm. Colt automatic guns. The accuracy of the fire from these was testified to by our prisoners.

21. Assistant Engineer Procter has just joined the ship. During the battle he acted as your aid, and afterwards took charge of a boat, with which, at great peril to himself, he saved many lives from the Pluton and Furor.

22. Assistant Surgeon Bransford did double duty. He took charge of a gun in Mr. Edson's division and fought gallantly through the action, his services as surgeon not being called upon until the wounded prisoners were brought on board.

23. Among the men, I beg leave to call your attention to the services of John Bond, Chief Boatswain's Mate, Captain of No. 1 gun (3-pounder rapid-fire). The excellent record of this man on board ship is known to you. His remarkable marksmanship, perfect coolness in action, his control over men, and his force of character would indicate his fitness for a higher position than that he now occupies.

24.   William C. Bee, Chief Gunner's Mate, also deserves special mention at this time.   This man, an ex-apprentice, left a lucrative position on shore from a pure sense of duty and patriotism.   His constant services on board have been almost invaluable, and his behavior during the action and in the work of rescue under Ensign Edson should, in my opinion, be recognized by a material advancement that would keep him in the service.

25.   Herman C. Green, Quartermaster, 1st class, steered the ship through the action without an error, where an error could so easily be made with disastrous results.   His coolness and skill merit recognition.

26.   The division officers, whose reports to you accompany this letter, have made certain recommendations in which I heartily concur.

Very respectfully,
HARRY P. HUSE,
Lieutenant U. S. Navy, Executive Officer.

To the Captain.

LIEUTENANT T. C. WOOD.

U. S. S. Gloucester,
Off Santiago de Cuba,
July 5, 1898.

SIR:—

1. In accordance with instructions, I beg to submit the following report on our engagement of Sunday, July 3rd, with the Spanish fleet, so far as events came under my own observation and in my own experience.

2. At 9.40 A. M. the spar-deck divisions of the ship had been inspected as usual by the Commanding and Executive Officers and the word had been passed that they could leave their quarters, but should remain on deck. As the men of my division, 2nd or after division, were going forward, some of them called my attention to black smoke arising from just inside the entrance to Santiago Harbor. In a moment a vessel appeared followed closely by five others. These were immediately recognized as the ships comprising the Spanish fleet and it was at once realized that they were making a desperate attempt to escape the ships of our blockading squadron. As each of the enemy's ships reached the mouth of the harbor, it turned sharply to the westward and steered at full speed close in shore in the following order: Infanta Maria Teresa, Cristobal Colon, Viscaya, Almirante Oquendo, Pluton and Furor, their distance being about 1500 feet apart.

3. The battleships of our squadron were lying some three miles off Morro to the southward, the Brooklyn was well to the westward, and the flagship New York was away to the eastward (off Siboney). This ship (the Gloucester)

was occupying her usual day position a little to the eastward of Morro and about one and a half miles off shore. No other small vessels of our fleet were in sight.

4. As soon as the enemy appeared, our crew were immediately called to quarters, and the ship cleared for action, although time did not permit the more elaborate preparations for battle, such as covering boats with canvas or making any protection against splinters. The hose was led along, pump started, buckets filled, and all precautions taken against fire in my division. It was customary for each of the three guns of my battery to have provided always two full boxes of ammunition, each containing 16 fixed charges; these boxes were carried on deck near the guns. The battery of the 2nd division consisted of three 3-pounder rapid-fire Hotchkiss with cylindrical English mounts; the guns and mounts having been made by Whitworth of Elswick, England. One gun was mounted directly aft, amidships on the overhang of the stern, and had an arc of train of about 220°. Similar guns were mounted, one on each side, thirty feet forward of the stern, and each possesses an arc of train of about 100°. The charges consist of armor-piercing shell fixed in brass cases containing cordite or smokeless powder.

5. The order to clear ship for action was at once followed from the bridge by an order to " commence firing "; and, as the ship lay with her stern towards Morro, I was able to put my battery in action immediately, the first shot being fired by the stern gun at the leading Spanish cruiser at a range of 3800 yards, the time being 9.43 A. M. My starboard gun followed immediately, and both continued rapid fire until the ship, by heading in straight for the enemy, under a strong port helm, threw them out of action and brought the guns of the forward division to bear. As the Spanish line of battle became more extended and the

two torpedo-boat destroyers, Pluton (leading) and Furor, appeared at the end of the line, we started a converging course and again brought my starboard and stern guns to bear. On the appearance of the destroyers we brought our full fire on them and so continued to the end of the action, leaving the enemy's cruisers to the battleships Indiana, Iowa, Oregon, Texas and the cruiser Brooklyn, which gradually closed on them. The destroyers speeding westward replied to our fire, but with deliberation and great inaccuracy. As we were within range of Morro and the other batteries, I noted the shells of their guns falling near us. During the action the machine gun of the Furor was trained on us and I noted the shots striking the water and gradually reaching towards us as the range decreased. The fire suddenly stopped before the bullets actually found us, owing probably to the disconcerting effect of our fire upon that vessel. As we approached the enemy our range was reduced, the last shots from my division being fired at 200 yards. At 600 yards I observed the leading destroyer (Pluton) was heading towards and close to shore about two miles to the westward of Morro. Smoke and steam were rising from her, her fire was almost stopped, and she was quite a distance in. Her companion ship, the Furor, was astern, but closing her distance and with a starboard helm. She was pointing off shore and in our direction, leading me to think she was about to close with us. She continued to turn, however, completing her circle to port at decreasing speed. Steam and smoke from her decks and her slackening fire showed her to be damaged, and it was evident she was disabled too seriously to continue the action. Soon a white flag waved amidships indicated her surrender and I received permission from Captain Wainwright to board her in the hope of extinguishing her fire and saving her. The dinghy manned by Collin and Thomp-

son (seamen), John Bond, Chief Boatswain's Mate, and
Lykke (seaman), accompanied me. On reaching the Furor
a scene of horror and wreck confronted us; the ship was
riddled by three and six pound shells, although I observed
no damage by large projectiles; she was on fire below from
stem to stern, and on her spar deck were the dead and hor-
ribly mangled bodies of some twenty of her crew and officers.
One of her boats was at the davits, smashed to atoms; another I afterwards found a short distance away, stove but
sustaining two survivors, whom I rescued. In the meantime another of the Gloucester's boats arrived and boarded
the wreck in charge of Lieutenant Norman, and between us
we rescued some ten or twelve of the crew which remained
on board. Finding it impossible to save the ship and fearing damage to our own men from explosions, I directed our
two crews with the surviving members of the Furor's crew,
to instantly abandon the ship and return to the Gloucester.
This was done and I was so fortunate as to find and take
with me the Furor's ensign. An explosion occurred almost
immediately after we abandoned the vessel, and her stern
commenced slowly to settle for ten minutes when her bow
began to rise. When the ship was nearly up and down she
sank about two hundred yards from shore. On reaching
the Gloucester I was directed to take the whaleboat and
go ashore for survivors, as in the meantime the Pluton was
on the rocks and her crew could be seen in the water and
on shore. On this trip, with Jaggi, Coxswain, we succeeded in bringing from shore the Captain and one or two
officers of the Furor, together with some twelve other survivors of both destroyers. The Gloucester in the meanwhile had proceeded to the westward to engage in the rescue of men from two of the Spanish cruisers.

6. After putting my prisoners aboard, I was ordered to
save what life I could from the Almirante Oquendo, hard

and fast ashore and burning furiously. This vessel lay with her bows in shore almost perpendicular to the beach and some three hundred yards from it. On going alongside as near as practicable owing to the surf and great heat from the burning vessel, I could see none of her officers or crew except some twenty or thirty crowded on the forecastle or hanging by ropes from her bows, and these I succeeded in rescuing and putting aboard our ship, together with some ten or twelve which I found floating on fragments of wreck. The burning cruiser, her plates many of them burst outboard and red hot, the roar of the flames, the constant explosion of small arm ammunition varied by heavier explosions from the guns or boilers, this with the cries of the wretches on her bows for help all made a scene which was indescribably impressive. The gratitude of those we saved was touching and from all the Spanish survivors I saw only a devout appreciation of their escape and from many of them a testimony of the terrible effectiveness of the fire from our fleet.

7. In the action I commend heartily the Captain of Guns' Crews Nos. 9 and 10, Adam Jaggi and F. W. Lacy, whose coolness and accuracy were admirable.

8. Among the crews of the guns I saw only enthusiasm and a remarkable indifference to the enemy's fire.

9. Beginning the action in their mustering clothes, the men stripped to the waist, and at the end were as fresh and eager to handle their guns as in the beginning.

10. I can commend heartily the guns (3-pounder quick-firing Hotchkiss). About 432 shell were expended in the battery under my command in the engagement, which lasted from 9.43 A. M. to 11 A. M. In two instances the charges failed to explode and were withdrawn; but in no case was there a jam or the slightest accident or delay in working the guns.

11. The conduct of the boat's crews in making the rescue was excellent; though under great excitement, and personal risk, the men acted coolly and obeyed orders promptly; the more remarkable considering the short time the Gloucester has been in commission.

12. Many stirring scenes were enacted, especially in the work of rescue, and much individual heroism displayed on the part of the crew which would make this report too lengthy to relate, but which fully maintain the best traditions of the American Navy.

Respectfully submitted,

THOS. C. WOOD,

Lieutenant U. S. Navy.

To Lieutenant-Commander Richard Wainwright, U. S. N.,
Commanding U. S. S. Gloucester.

LIEUTENANT G. H. NORMAN.

U. S. S. Gloucester,
Off Santiago de Cuba,
July 6, 1898.

SIR:—

I have the honor to report that on the morning of July 3rd, after the sinking of the Spanish torpedo-boat destroyer Pluton and crippling of the destroyed Furor, I went in charge of the first whaleboat to the rescue of the living on the Furor, having as crew the following men: Jaggi (cox.), Evans, Quentin, Tierney, Daley, Loehrs, Lawerence and Rozzle.

On approaching the Furor I could see that her sides were riddled with shot holes, ranging in size from those made by rifle bullets to some three inches in diameter. These and a rent in her starboard freeboard a foot or two long and from an inch to three inches in width comprised all the damage which her outer hull seemed to have suffered during the action. I, with several of my men, climbed upon her decks. Upon these there were lying many dead, and many more could be seen through the hatches in the spaces below. The vessel was afire in every part; and I withdrew with our boat, carrying off with me all the living of the Furor's crew then aboard her, eight in number, and returned on board the Gloucester. After the Gloucester had steamed to the eastward to a point several miles beyond where we had driven the destroyer Pluton on the rocks, I went away in charge of the gig to the rescue of the crew of the Infanta Maria Teresa who could be seen crowded upon the bows of their ship, the after part and waist being afire

and burning fiercely.   I had as crew of the gig Dahl (cox.),
Evans, Thompson, Magee, Kastell and Cooksey.   The
Teresa had run aground and lay one hundred and fifty yards
from shore.   As I approached her, I could see some of her
crew, about a dozen, already on the beach, surrounded by
a little band of Cubans.   Mr. Edson, in charge of another
of our boats, having carried a line from the bow of the
Teresa to the shore, we immediately set about disembarking
her crew; letting those that were badly wounded be lowered
by ropes to our boats, but compelling the uninjured ones
to come down and out on the life line until they could drop
into one of our two boats, which we kept a few yards from
the ship's sides.   By using one of our boats to receive the
men and the other to ferry them to the surf, we got ahead
rapidly, and in less than three hours had landed all the living
from the ship, to the number of four hundred and eighty.
Of these, many were wounded; but they and all the rest had
to be put over in the water when forty yards from the shore
and dragged through the surf to the beach.   I rescued in
the first boat-load from the decks of the Teresa a Spanish
officer who could speak some English.   By retaining him in
my boat I was able in some measure to direct the actions of
those on the ship.   Through him I received the promise
of the officers set on shore that they would, so many of them
as I wished, return with me to the Gloucester, as soon as
our work of rescue was finished.   All through the time we
were rescuing the crew of the Teresa, small explosions were
constantly occurring on and between her decks.   The fire
was steadily working forward, and those still left in the
ship were urging us to hurry in our work of removing them
as they feared an explosion of the forward magazine.   After
the crew of the Teresa had been gotten ashore, I backed
my boat in on the life line as near to the surf as possible and
sent a man ashore with orders for the Admiral, the Fleet

Captain, and the five other officers next in rank, to come out to my boat, which they promptly obeyed, two of our own men dragging them one at a time along the life line through the surf to our boat's side. I then returned with these, and the officer whom I had kept with me throughout as interpreter, as prisoners to the Gloucester. Throughout the long pull out, Admiral Cervera and his officers expressed much gratitude for our rescue of them and their crew and considerable anxiety for the safety of those we had left on the beach, who though over four hundred in number, being unarmed as well as nearly naked, were at the mercy of the Cubans who had gathered about them.

As soon after my return with my prisoners to the ship as the cutter could be provisioned, I was again sent away with an armed party consisting of Bond (cox.), Thompson, Noble, Collin, Mulcahey, Lykke, Rozzle, Lewis, Halverson, Hillman, Dahl, Tierney and Loehrs, to the rescue from the Cubans of the Spanish whom we had earlier taken off the ship. On reaching the shore, I found that the Spanish, fearing the explosion of their ship's magazine, had retreated from the water-front and were gathered behind the shoulder of a hill which reached down to the eastern end of the beach. Some few, however, were still on the beach carrying the wounded to a more sheltered place. There were also half a dozen dead lying in the sand who had succumbed from their wounds after being gotten ashore. They were later buried. Having landed my stores and placed over the the boat an armed sentry, I, with a guard and the colors, proceeded to where the main body of the Spanish had gathered and found a small band of armed Cubans watching over them. I then, in the presence of these Cubans, had the Spanish formally surrender to me through their Commanding Officers, the third officer of the Infanta Maria Teresa and one from the Oquendo, many of whose crew had by that time straggled over from where the latter ship lay wrecked a mile further down the coast. The Cubans were then informed through one of the Spanish officers as an interpreter that I, with my armed party, would take exclusive charge of my prisoners; and the Cubans withdrew, threatening, I was told by the Spanish Officers, to come back in force. I then went out to the beach, and the U. S. S. Harvard appearing in the offing, I sent her a wig-wag message that I had five hundred prisoners in need of provisions and protection. She sent a boat ashore, and on its return to the Harvard with my report, she sent three other boats; and the work of transferring the Spaniards from the shore to the Harvard immediately began. The Indiana also sent two surgeons, a steam launch with a pulling boat in tow, and medical supplies. In this boat and our cutter the surviving of the wounded, after being attended by the Surgeons, were carried to the Harvard's, the boats being beached to permit of getting the wounded out through the surf. The uninjured Spanish were sent, sometimes dragged, through the surf, which by this time, however, had fallen somewhat with the wind, and packed into the boats of the Harvard which carried them out to that ship; the last boat-load leaving the beach at some time after ten o'clock at night. A little later, the Gloucester's boat having got away with the last of the wounded, I, with those of the Gloucester's party left ashore, went in one of the Harvard's boats to the Harvard. Our cutter was hoisted upon her deck and my party taken on board, fed, and given clothing. The following morning at about eight o'clock we were lowered out off Siboney and returned to the Gloucester which was lying alongside.

In these three expeditions of July 3rd, the conduct of the several boats' crews was in each case most creditable. Without the coolness and unflagging zeal which they displayed, the rescue of a large portion of those we saved and the success of the work assigned to me would have been impossible. I feel that I should make particular mention of the good work done by Chief Boatswain's Mate Bond, Boatswain's Mates Dahl and Thompson, Quartermasters Evans and Noble, and Loehrs, Lykke, Hillman, Mulcahey, Collin, Rozzle, and Halverson. Of these I beg leave to report as especially deserving of commendation, Bond, Dahl, Thompson, Noble and Jaggi; their general bearing and the example they set the other men being worthy of the highest praise.

Of the action of the morning of July 3rd, I have to report that throughout the engagement the three guns in my division, Nos. 1, 3 and 5, worked satisfactorily, with the exception that the shield bolts of both six-pounders were broken, requiring later the removal of the shields.

The crews of these guns displayed good courage and coolness during the action. Their prompt obedience to orders and steadiness under fire cannot be too much praised. The good work done by the bow gun, No. 1, Bond, Captain, deserves especial mention; as does also the good gunnery of Dahl, Captain of No. 5 gun, and Whitelock, Captain of No. 3 gun. Of others in the crews serving these guns, I wish to report for previous interestedness and pains taken in the care of the guns in their charge: Thompson, Mulcahey, Harbour, Kastell and Hillman.

Very respectfully,
George H. Norman, Jr.,
Lieutenant U. S. N.

To Lieutenant-Commander Richard Wainwright,
Commanding U. S. S. Gloucester.

ENSIGN J. T. EDSON.

U. S. S. Gloucester,
Off Santiago de Cuba,
July 4, 1898.

Sir:—

I have the honor to report on the conduct of my division in the action of July 3rd, during which the Gloucester sank the torpedo destroyers Furor and Pluton and rescued four hundred and eighty officers and men of the enemy's fleet.

During the engagement, my men worked willingly and well, and I observed not one instance of cowardice or confusion. The powder division, assisted by guns crews 2 and 4, whipped up most of the ammunition early in the engagement while the starboard guns were in rapid action. When my guns 2 and 4 were brought to bear on the enemy, their fire was as good as that shown at target practice. And it was during this fire that the boilers of the Pluton were pierced. Peter Keller, Boatswain's Mate, Patrick Lynch, Coxswain, and myself did the firing at these guns.

The sighting of No. 4 gun was rendered difficult by reason of a weak spring in the cylinder. The spring did not fully return the gun after its recoil, thus throwing additional weight on the man who sighted it. In firing this gun I found it necessary to brace my body firmly against the shoulder piece and I was unable to do good shooting for more than a few rounds at a time.

In the heat of the action, the firing pin of No. 2 dropped out while Keller was shooting. The breech-block was instantly removed, the pin found and replaced by Gunner's Mate William C. Bee and his assistant M. J. Murphy. The time occupied seemed to me about two minutes. I con-

11

sider such cool, intelligent action in the midst of a rain of
shot and shell worthy of commendation.

The bolts of No. 5 gun, which held the gun shield, were
fractured during the engagement, but the shield was not
removed until we had ceased firing. This damaged shield
was a slight menace to the men operating this gun, but it so
happened that I fired the last dozen shots from this gun by
the order of Captain Wainwright.

After the Furor had been sunk and the Pluton had exploded, I was sent with my division in command of the
ship's cutter with instructions to rescue the crew of the
Infanta Maria Teresa. As we approached this vessel, which
was in flames, I saw the crew crowded forward on the forecastle and I noted that the vessel lay nearly broadside on
to a sandy beach distant about two hundred yards. As we
neared her, I held up a rope's end to indicate my purpose.
A line which they gave me I took to the beach and called
for a good swimmer to take it through the surf. Otto
Brown responded so manfully to my call that it was easily
seen that he was the right man for the work. With the line
about his neck he fought against the breakers for twenty
minutes. He returned once to the boat for rest. The line
was more carefully tended after this by William C. Bee; and
after another struggle, the cutter being closer in this time,
he made the beach. I sent Keller also through the surf to
secure the line ashore. The cutter was hauled to and from
the ship along this line, carrying each time eight or ten men
from the burning wreck. As we neared the beach each
time, I found it necessary to throw one or more Spaniards
into the water in order to expedite the work. These men
were immediately grabbed by Keller or Brown and passed
along the line to the beach. In this manner the cutter
landed about two hundred officers and men, and I believe
that Admiral Cervera was among the number. The only

other boat engaged in this rescue was the gig from the Gloucester in charge of Lieutenant Norman. We worked frequently in association and succeeded in rescuing every soul on board without losing a single life. Among the rescued were many wounded who required much assistance. This was ably rendered by Keller and Brown, who remained in the water for about two hours.

The cutter also rescued men from the ship's cables and from boatfalls amidships, but not until every man on the wreck understood that my will was for them to come over the line which had been run. By following this latter method I could absolutely control the number of men carried each trip, and the position of my boat was more favorable in the event of any serious explosion. During the whole time that the men engaged in the work, there were occasional reports on board as from exploding ammunition. As we left the wreck for the Gloucester, the fire had reached the forward turrets so that the whole ship was nearly aflame. One of the eleven-inch guns was discharged, sending a shot into the water close alongside of us.

During the engagement Dr. Bransford aided me in the command of my division; and after the fight and the rescue, my knowledge of surgery enabled me to render him assistance in caring for the wounded.

The men who went with me in the cutter were as follows: Peter Keller, Boatswain's Mate, 1st Class; Otto Brown, Seaman; William C. Bee, Gunner's Mate; W. H. Sellers, Engineer's Yeoman; Milo K. Davis, Landsman; Oscar Halverson, O. Seaman; P. Kay, Ship's Cook; J. W. Lewis, O. Seaman; Griffith Roberts, O. Seaman; Neil H. Lykke, Seaman Gunner.   Very respectfully,

J. T. EDSON,
Acting Ensign, U. S. N.

Lieutenant-Commander Richard Wainwright, U. S. N.,
Commanding U. S. S. Gloucester.

CHIEF-ENGINEER G. W. McELROY.

U. S. S. Gloucester,
Off Santiago de Cuba,
July 4, 1898.

SIR:—

In obedience to an order received to-day, I have the honor to make the following report concerning the work done in the Engineer Department, and the conduct of the men of the Engineer Division, during the engagement of our fleet with the Spanish fleet on the 3rd instant.

The machinery worked well, no trouble arising from any cause. Forced draught was used for a time, and it is estimated from the revolutions made that a speed of 17 knots was attained.

As no shots entered the boiler or engine compartment, and as there was no accident to the machinery, there was no opportunity for signal instances of heroism or display of special presence of mind. The men behaved splendidly from the Chief Machinist down, showing the strictest attention to duty, and an anxious care that nothing should go wrong at such a critical time.

2. A considerable number of the force are new to the service, having enlisted from a sense of duty. As a fitting recognition of the great service rendered by them, I respectfully recommend the following changes in ratings, and take pleasure in testifying to the worthiness of each man named:

1. The acting appointment of Robert Jennings to be made permanent and that he be held in favorable remembrance should the grade of Warrant Machinist be established.

2. That Carl Johanson, holding an acting appointment of 1st Class Machinist, be made a Chief Machinist.

3. That Robert J. Hare, 2nd Class Machinist, be made 1st Class Machinist.

4. That M. J. Carabine, 2nd Class Machinist, be made 1st Class Machinist.

5. That T. S. Barnes, C. E. Woodside, and B. Bowie, 2nd Class Machinists, be made 1st Class Machinists. They have not yet had much handling of marine engines, and acting appointments instead of permanent ones are therefore recommended. They handle the engines of this ship, whenever called on, most capably, and only need a little practice to qualify them in every way.

6. That H. Roberts, 1st Class Fireman, be given an acting appointment at Water Tender.

7. That James McMillan, 1st Class Fireman, be made Acting Oiler.

8. That Harry McNabb, 1st Class Fireman, be made an Acting Oiler.

9. That E. C. Adkins, Thomas Colleran, and Andrew Cole, 2nd Class Firemen, be made 1st Class Firemen.

10. That W. J. Sullivan, E. Howard, P. Welch, and Walter Lee, Coal Passers, be made Firemen, 2nd Class.

3. All other men in the force on duty, while doing most excellent work and therefore recommended to your favorable consideration, are from the nature of their ratings, their physical condition, and personal reasons, not at present qualified for promotion. Notable among the first class are Chief Yeoman W. H. Sellers, Blacksmith W. A. McArthur and Acting Water Tender W. H. McKeon. The last-named man is a most capable man. He tended water during the engagement. I regret that, from lack of shop knowledge, he cannot be recommended for a Machinist's billet. He holds a permanent appointment as Oiler, which he desires

to resume when his acting appointment as Water Tender expires.

4. The following men were not on duty during the engagement: 2nd Class Machinist George Rudischauser. sick; Coal Passer G. S. Veneau, confined.

Very respectfully,
GEORGE W. McELROY,
P. A. Engineer. U. S. N.

The Commanding Officer,
    U. S. S. Gloucester.

REPORTS OF

CAPTAIN HIGGINSON, CAPTAIN WAINWRIGHT

AND LIEUTENANT HUSE ON THE CAPTURE

OF GUANICA, JULY 27, 1898

LANDING PARTY* AT GUANICA, July 25th, 1898.

Lieutenant Huse, Commanding.
Lieutenant Wood,
Lieutenant Norman,
Assistant Engineer Procter,
Assistant Paymaster Brown.

Lacy,	Whitelock,	Bechtold,	Chipman,
Mulcahey,	Tierney,	Magee,	Keller.
Murphy,	Wirtane,	Collin,	
Loehrs,	Davis,	Thompson,	
Macklin,	Harbor,	Kroning,	
Rozzle,	Kay,	Hillman,	
Kastell,	Halverson,	Butler,	
Englert.	Carlsson,	Buchanan,	
Roberts,	Kleinkopf,	Lawerence,	

* This roll includes the reenforcements sent from the Gloucester after the enemy opened fire.

CAPTAIN HIGGINSON.

[By cable from St. Thomas, July 26, 1898.]

To Secretary of Navy, Washington.

Arrived here with General Miles and the United States troops to-day at 9.15 A. M. Landed them safely. There are no batteries outside. Gloucester entered the harbor, landed company of sailors under command of Lieutenant Harry P. Huse, U. S. N., and Lieutenant Wood. Dispersed small company of Spanish soldiers. Hoisted flag. Commend Lieutenant-Commander Richard Wainwright and officers for gallantry capturing Guanica. Transport discharged without any opposition, assisted by boats of Massachusetts.

I have telegraphed Admiral Sampson.

HIGGINSON.

Port Guanica, Puerto Rico, July 25.

CAPTAIN WAINWRIGHT.

U. S. S. Gloucester,
Off Arroyo, Puerto Rico,
August 4, 1898.

SIR:—

I have the honor to forward herewith the report of Lieutenant Huse, who commanded our shore party at the seizure of Guanica.

We entered the harbor by permission of the Senior Officer Present, and fired at some fleeing troops, then landed a party to seize the available landing places, and prevent the destruction of lighters.

Reinforcements were discovered coming from Yauco, but were driven back by fire from this vessel.

The army transports came in sight with launches and boats from the vessels in the outer harbor. At my request, Col. Black immediately landed a portion of his engineer battalion and the village was turned over to the army.

General Miles visited the Gloucester and thanked us for the services rendered.

The landing party was well handled by Lieutenant Huse and the men behaved extremely well, particularly when it is remembered that it was their first experience on shore.

The behavior of the Navy rifle was due, in my opinion, partly to lack of training, and partly to mechanical defects that can be remedied in later models.

Very respectfully,

RICHARD WAINWRIGHT,
Lt.-Comdr., U. S. N., Comdg.

The Commander-in-Chief, Naval Forces,
North Atlantic Station.

LIEUTENANT HUSE.

U. S. S. Gloucester,
Guanica, Puerto Rico,
July 25, 1898.

SIR:—

I have the honor to report as follows on the landing party
sent this morning to capture the village of Guanica:

The force under my command consisted of twenty-eight
men and Lieutenant Wood, embarked in the cutter. We
landed, without meeting with any opposition, at a little
wharf, and the men were at once deployed to cover the
beach. The Spanish flag was hauled down and our colors
hoisted in their place.

This drew the enemy's fire, who opened from the under-
brush on the right flank and from about three hundred
yards distance on the highway. We answered with a slow
fire from the skirmish line. The cutter took up a position
at the foot of the highway that leads up into the country,
and fired short volleys from the 6-millimetre automatic
Colt, but this soon jammed and was of no further use in the
action. I sent Lieutenant Wood to take charge of the right
flank with eight men, while Chief Yeoman Lacy with four
men thoroughly covered the left flank from the ruins of a
stone house well situated for the purpose. From a country-
man, the only man in the village, I learned that we were
opposed by thirty regulars, and that reinforcements were
momentarily expected from Yauco, about four miles distant.
I signalled to you for reinforcements and pushed forward
our center along the highway. The enemy's fire was well
sustained but high, and no casualties resulted from it. At

the northern limit of the village we built a wall across the highway and placed there the new Colt gun you had sent ashore. We also strung two barbed wire fences fifty and one hundred yards to the front across the road. Meanwhile a boat under command of Assistant Engineer Procter was engaged in cutting out a large lighter, which came into immediate use in landing troops.

About this time the Gloucester opened fire from her 3-pounders and 6-pounders and the enemy retreated. A few minutes later the first contingent of the regular army, Colonel Black's regiment of engineers, landed and rapidly pushed forward beyond our lines. In obedience to your orders the landing party then returned to the ship. At the special request of General Gilmore I left Lieutenant Wood and a party ashore with the Colt gun. They returned to the ship an hour later.

I wish especially to mention the gallant conduct of Lieutenant Wood and of Chief Yeoman Lacy. All the men under my command behaved splendidly. The Navy rifle behaved abominably, the majority jamming at inopportune moments, and several being rendered useless when we appeared to be in considerable danger of defeat.

Very respectfully.

HARRY P. HUSE,
Lieutenant and Executive Officer.

To the Captain.

REPORTS OF
CAPTAIN GOODRICH AND LIEUTENANT WOOD ON
THE CAPTURE OF ARROYO, PUERTO RICO,
AUGUST 1, 1898

LANDING PARTY AT ARROYO, PUERTO RICO, AUGUST 1, 1898.

Lieutenant Thomas C. Wood, Commanding.
Lieutenant George H. Norman, Jr.
Assistant Paymaster Alexander Brown.
Apothecary Beckary.

Lacy,	Whitelock,	Bechtold,	Chipman,
Keller,	Thompson,	Collin,	Englert,
Jaggi,	Harbour,	Murphy,	Buchanan,
Butler,	McNab,	Quentin,	Noble.
Riley,	Wirtane,	Davis,	
Roberts,	Magee,	Mulcahey,	
Lynch,	Rozzle,	Macklin,	
Tierney,	Hillman,	Kastell,	
Loehrs,	Lawerence.	Kleinkopf,	

CAPTAIN GOODRICH.

U. S. S. St. Louis,
Arroyo, Puerto Rico,
August 2, 1898.

SIR:—

I have the honor to report receiving the peaceful surrender or Arroyo, Puerto Rico, yesterday at the hands of the Alcalde under the following conditions:

2. a. The civil authorities to remain in power during the pleasure of the Government.

b. The religious authorities to exercise their influence toward suppressing disorder and maintaining peace.

c. All the lighters in port, five in number, to be placed at the disposal of the United States with crews of natives.

d. All Spanish Government property and papers to be surrendered.

e. The lighthouses to be maintained by the former keeper, his pay to continue as before, at the charge of the United States.

These conditions were agreed to by the Alcalde, the Judge, and the Parish Priest.

3. At 11.25 A. M., August 1st, the American flag was hoisted over the Custom House, and a guard of enlisted men from the Gloucester, under Lieut. Wood of that vessel, placed in charge of the town.

4. I was personally on board the Gloucester at the time, having been sent by the Senior Officer Present at Ponce, Capt. Ludlow, U. S. N., with her and the Wasp to reconnoiter Arroyo and Jobos with reference to landing Major-Gen. Brooke and his staff and the 1st Army Corps.

5. The details of the occupation were left to Lieutenant-Commander Richard Wainwright, to whose judgment and advice I am greatly indebted. Lieutenant Ward of the Wasp after accompanying the Gloucester to the anchorage, prepared to lend assistance in the event of opposition, rendered valuable service pending the negotiations by running lines of soundings along which deep draught vessels might safely approach. Upon his work was based the adoption of Arroyo as the landing place.

I am sir,

Very respectfully,

C. F. GOODRICH,

Captain, U. S. N., Commanding.

The Commander-in-Chief, N. A. Station,
    U. S. F. S. New York.

LIEUTENANT THOMAS C. WOOD.

U. S. S. Gloucester,
Off Arroyo, Puerto Rico,
August 3, 1898.

SIR:—

I beg to make the following report upon the capture and occupation of Arroyo, Puerto Rico, by the landing party from this ship on Monday, August 1, 1898:

At 9.30 A. M., the ship being anchored in the open roadstead about three-quarters of a mile from shore and opposite the town, you directed me to go ashore in the whaleboat under a flag of truce, gather the officials of the place, and demand its surrender to the United States in the name of Captain Goodrich, U. S. N., Senior Officer Present, then aboard this ship. I was also instructed to recommence the light on Figuera Point; and further, to obtain possession or use of all the lighters in the harbor for the purpose of landing troops and stores when they should arrive in the expected transports. On this errand I took with me, with your permission, Assistant Surgeon John F. Bransford to act as interpreter.

On approaching shore and finding no landing, I beached our boat. Assembled on the beach was a large crowd of natives and a few police with side arms. In front and apart from these stood a Spanish officer in uniform to whom I addressed an inquiry for the Commandant and town officials. He conducted us to the Custom House fronting the water, and in a few moments were gathered himself, the Captain of the Port, the Alcalde, the Collector and his Deputy, the Justice of the Peace, the Padre, and some of

12

the leading citizens. Through Dr. Bransford I informed
these gentlemen of the presence of the Gloucester and demanded the immediate surrender of the town and all Spanish property. A heated discussion followed which I was
obliged to cut short by a peremptory request for an affir-
mative answer under penalty of a prompt bombardment of
the town. One by one the gentlemen yielded their surrender and gave their parole, with the exception of the
Captain of the Port, a Spanish naval officer (retired) who
refused to do either. All these gentlemen I then sent off
to the ship in charge of Dr. Bransford, whom I requested
to report to you the condition of affairs. I kept ashore
with me Chief Quartermaster Bechtold and then directed
him to hoist the U. S. Flag on the Custom House. This
was done, and our colors were greeted with cheers from
many of the natives and negroes.

At about 10.30 o'clock a landing party from the Gloucester arrived, consisting of 35 men with a Colt machine
gun, all under command of Lieutenant Norman and Assistant Paymaster Brown, with orders to report to me for
the occupation of the town. Pickets were immediately
placed in the principal streets, Lieutenant Norman taking
charge of the right wing on the eastern side of town, Paymaster Brown with the Colt gun commanding our left or
western wing, including the main road to Guayama (three
miles distant), while I held the center with the Custom
House (opposite the landing) as Headquarters and Guard
House. Orders were issued to allow unarmed and peaceful citizens to pass freely about and into town, to put under
arrest all suspicious and armed persons, and to prevent the
exit of any one beyond the town limits. Some exceptions
were made to the latter rule in the case of English and
French residents whose homes lay beyond the town lines,
and passes were issued to these persons. Our men were

strictly enjoined to respect the property and persons of citizens, and the latter were warned that any treachery or any opposition would be instantly and severely punished. Using excellent discretion, Lieutenant Norman seized and held the telegraph office and receiving instruments, cutting the lines to prevent information reaching the enemy. Through various channels we were in constant receipt of information of the presence on the outskirts of town of bands of Spanish guerillas and Guardia Civile, the most reliable pointing to some sixty or eighty encamped on the Guayama road, while several hundred were said to be holding that city. No attack was made on Arroyo, however, during our occupancy.

At about 1.30 P. M., Dr. Bransford returned on shore bringing the party of Spanish officials whom I had sent to the ship. As the Captain of the Port had decided to yield to superior force and give his parole under compulsion, I was directed to receive this and set him at liberty with the privilege of retaining his sword. Dr. Bransford then returned to the ship carrying with him my verbal report to you of the disposition of our forces.

During the day, arrangements had been made with the Alcalde to recommence the light on Figuera Point; and at sunset it was burning in accordance with his promise and has been kept alight every night since. Inquiries as to lighters showed some eight or nine of them drawn up in shore; most of them being the property of Mr. McCormick, former U. S. Consul. Mr. McCormick promptly promised the use of these lighters and proceeded to set them afloat and anchored them near the Gloucester.

During the afternoon, signals were made to withdraw our force at dark. Friendly residents were notified and offered an asylum aboard the Gloucester. Mr. McCormick, fearing an attack on his house in the country by the guerillas, brought his family and the families of his brothers into town

under the guns of the ship; and these precautions proved
to be wise ones, since during the night a party of the enemy
raided the town and fired a few shots at the ship and retreated, although the ship's searchlight was kept on the
shore.

At dawn on the following day (Thursday the 2d inst.), the
same force and under the same officers was sent ashore to
re-occupy the town, and the day passed much as the previous
one had done. We noticed, however, that the attitude of
the people was rather more unfriendly; accounted for, we
thought, by their doubts as to the strength of our force,
and our intentions. To be sure, the St. Louis had arrived
at daylight; but information as to the weakness of our
party was in possession of the enemy, and the country people brought us word that a troop of mounted men was on
its way to attack us. I was just about to ask for a few shells
to be fired from the Gloucester when you signalled to ask in
what direction you should point your guns. The firing
which followed and the shells which you dropped on our
left flank I have no doubt prevented this threatened attack.

In the afternoon we were glad to see the troops coming
to our relief from the St. Louis. By five o'clock some two
or three hundred had been landed; and in accordance with
your orders, I turned over my command to the first officer
of rank who came ashore, Colonel Bennett, of the 3d Regiment Illinois Volunteers, to whom I carefully explained the
situation and introduced formally to him the officials of the
town. Colonel Bennett then detailed Lieutenant Hayes and
some sixty of his men to relieve my pickets; and the latter
with Lieutenant Norman and Paymaster Brown, were withdrawn to our boats and returned to the ship at 5.30 P. M.
In the afternoon, I had detailed Lieutenant Norman to
make an inventory of the money and valuables in the
custom-house safe. It was done by this officer in a thor-
ough and businesslike manner, receipts were given to the

Spanish Collector of Customs, and, when Colonel Bennett took control, he in turn receipted to Lieutenant Norman.

As to the *personnel* and arms of our force, this report would be incomplete if it did not testify to the conduct of both officers and men. Although exposed to intense heat and to drenching rains for nearly two days, with little food, and without any relief on picket duty, the behavior of the force reflected credit upon the ship and the naval service. Lieutenant Norman and Paymaster Brown were alert, intelligent, and to them both I owe many valuable suggestions and the greatest help and support. Chief Yeoman Lacy, an old soldier, I made Sergeant of the Guard. He reported to me regularly the condition of the pickets and made his rounds every hour. In addition to this, he made a scouting trip alone to a distance of about half a mile from town in all directions. Chief Quartermaster Bechtold, Paymaster's Yeoman Whitelock, Apothecary Beckary, Boatswain's Mates Keller and Thompson, Coxswains Jaggi and Lynch, Gunner's Mate Collin and Quartermaster Noble are all petty officers whose examples to the crew did much to maintain the *morale* of our force. The following is a list of the enlisted men engaged in the expedition: Quentin, Loehrs, Mulcahey, Kastell, Butler, Buchanan, Tierney, Lawerence, Harbour, Macklin, Rozzle, Magee, Wirtane, Murphy, Hillman, Kleinkopf, Riley, McNab, Roberts, Davis, Chipman, Englert.

These men were armed with the new Navy Lee magazine rifle which on this occasion gave us no trouble. The men were equipped with cartridge belts, canteens, and haversacks, and their uniform was white working clothes, white hats and the regulation navy leggings.

Respectfully submitted,

Thomas C. Wood,

Lieutenant, U. S. N.

Lieutenant-Commander Richard Wainwright, U. S. N.,
Commanding U. S. S. Gloucester.

Endorsement by Captain Wainwright:

This report was not forwarded to the Commander-in-Chief as the transaction took place under the eyes of Captain C. F. Goodrich, U. S. N., commanding U. S. S. St. Louis.

The St. Louis anchored in the morning some miles from shore at a point selected as a safe anchorage the previous day. The Gloucester was expected to steam out and take the boats containing the troops in tow. Because of the threatened attack, it was deemed unwise to withdraw the Gloucester from her close proximity to the shore. This caused the delay in landing the troops. When the Gloucester opened fire, Captain Goodrich at once steamed the St. Louis towards shore until her bow touched the reef, opening fire with her battery. This at once settled the question of our being attacked by the enemy.

RICHARD WAINWRIGHT,
Lt.-Comdr., U. S. Navy, Comdg. U. S. S. Gloucester.